Safe Utilization and Ecological Restoration of Heavy Metal Polluted Farmland

Safe Utilization and Ecological Restoration of Heavy Metal Polluted Farmland

Guest Editors

Bin Guo
Ying Feng

MDPI

Basel • Beijing • Wuhan • Barcelona • Belgrade • Novi Sad • Cluj • Manchester

Guest Editors
Bin Guo
Institute of Environment, Resource, Soil and Fertilizer
Zhejiang Academy of Agricultural Sciences
Hangzhou
China

Ying Feng
College of Environmental and Resource Sciences
Zhejiang University
Hangzhou
China

Editorial Office
MDPI AG
Grosspeteranlage 5
4052 Basel, Switzerland

This is a reprint of the Special Issue, published open access by the journal *Toxics* (ISSN 2305-6304), freely accessible at: https://www.mdpi.com/journal/toxics/special_issues/0SD1NW1H74.

For citation purposes, cite each article independently as indicated on the article page online and as indicated below:

Lastname, A.A.; Lastname, B.B. Article Title. *Journal Name* **Year**, *Volume Number*, Page Range.

ISBN 978-3-7258-7985-4 (Hbk)
ISBN 978-3-7258-7986-1 (PDF)
https://doi.org/10.3390/books978-3-7258-7986-1

Contents

toxics

MDPI

Editorial

Safe Utilization and Ecological Restoration of Heavy Metal Polluted Farmland: Latest Strategies for Remediation

Bin Guo [1,*] and Ying Feng [2]

1 State Key Laboratory for Quality and Safety of Agro-Products, Institute of Environment, Resource, Soil and Fertilizers, Zhejiang Academy of Agricultural Sciences, Hangzhou 310021, China
2 MOE Key Laboratory of Environment Remediation and Ecological Health, College of Environmental and Resource Sciences, Zhejiang University, Hangzhou 310058, China; yfeng@zju.edu.cn
* Correspondence: guob@zaas.ac.cn

1. Introduction

Heavy metal pollution of farmland has emerged as a pressing global environmental challenge, which threatens food security, ecological integrity, and human health [1]. Innovative strategies for the safe utilization of heavy metals and the ecological restoration of farmland are required to combat the persistence, mobility, and bioaccumulation of heavy metals such as cadmium (Cd), lead (Pb), and copper (Cu) in agricultural ecosystems [2]. This Special Issue consolidates eight original research articles that address key aspects of heavy metal polluted farmland management, covering contamination characterization, remediation technologies, microbial responses, and predictive modeling. The collected studies provide valuable insights into the complex interactions between heavy metals, soil, plants, and microorganisms, offering practical solutions for sustainable farmland management.

Over the past decades, research on heavy metal pollution control has developed from single-factor remediation to comprehensive ecological restoration. Traditional physical and chemical remediation methods are often costly and environmentally disruptive, highlighting the need for eco-friendly alternatives [3,4]. Phytoremediation, soil amendment, and microbial regulation have emerged as promising approaches, but their effectiveness varies with pollution types, soil properties, and regional conditions [5]. This Special Issue aims to consolidate cutting-edge research findings, identify knowledge gaps, and guide future directions in the field of heavy metal polluted farmland restoration.

2. Overview of Published Contributions

2.1. Contamination Characterization and Source Apportionment

Three studies focused on the temporal and spatial variation in heavy metal pollution and source identification, providing a scientific basis for targeted remediation.

Cao et al. (Contribution 1) investigated the temporal and spatial variation in eight toxic metals (As, Cd, Cr, Cu, Hg, Ni, Pb, Zn) in cultivated soil in Jiaxing, Zhejiang Province. The results demonstrated that compared with the early 1990s, Cd, Zn, Cu, Pb, Ni, and Cr concentrations increased by 86.2%, 27.2%, 41.1%, 45.6%, 4.94%, and 21.1%, respectively. Using local soil background values as the standard, all elements reached mild pollution levels, with Cd and Hg being the most severely contaminated. Industrialization, urbanization, and the use of metal-containing feed were identified as the main drivers of temporal variation, while soil types and crop cultivation methods dominated spatial differences.

Chen et al. (Contribution 2) employed principal component analysis (PCA) and positive matrix factorization (PMF) to identify sources of potentially toxic elements (PTEs)

 https://doi.org/10.3390/toxics14020154

in agricultural soils of Yingtan City, Jiangxi Province. The average concentrations of Zn, Cu, Pb, Mo, and Cd exceeded local background values, with Cu and Cd showing the highest pollution levels. Four main sources were identified: mining activities, natural sources, copper smelting, and agricultural practices. The integration of PCA-PMF with field surveys effectively pinpointed pollution sources and specifically highlighted the role of mixed fertilizer application in Cd and Cu accumulation.

Wang et al. (Contribution 3) proposed a modified PCA-MLRD model integrated with grouped principal component analysis (GPCA) and GeoDetector for quantitative source apportionment in the Chang-Zhu-Tan region, Hunan Province. The study found that Cd and As were the main over-standard elements, with Cd exceeding the standard by 1.8 times on average and with an over-standard rate of 90%. Metal manufacturing factories contributed 30–44% to Cd pollution and 42–58% to As pollution, while soil parent material was the primary source of Pb (37–61%). Pollution pathways— such as atmospheric deposition, irrigation water, traffic emissions, and irrigation water—showed a more pronounced effect within 1000 m. The modified model improved accuracy by accounting for pollution processes, providing technical support for precise pollution control.

2.2. Phytoremediation and Plant Tolerance

Three articles explored the potential of phytoremediation using different plant species, emphasizing the selection of high-efficiency remediation plants and optimization of remediation effects.

Lu et al. (Contribution 4) conducted a two-year field trial on 23 chrysanthemum cultivars in Cd-contaminated soil in Zhejiang Province. Six cultivars (including marigolds) showed high Cd accumulation (average >0.6 mg per plant), with rhizosphere soil remediation taking only 4–5 years. Fourteen cultivars exhibited good multiple-cropping characteristics, and five (e.g., QX-yz) demonstrated high heavy metal tolerance. The application of bamboo vinegar significantly promoted Cd absorption in chrysanthemums, and an economic benefit analysis showed that planting dominant cultivars could generate income slightly higher than the local average.

Xu et al. (Contribution 5) compared the Pb tolerance and accumulation capacity of two industrial hemp varieties (Yunma1 (YM) and Shaanxi Industrial Hemp (SM)) under Pb stress. Pb mainly accumulated in the roots of both varieties (70–80%), with YM showing twice the Pb accumulation capacity of SM at high concentrations (5000 mg/kg). YM retained 70% of absorbed Pb in roots and exhibited strong tolerance (tolerance index > 90%) even at Pb concentrations exceeding 4000 mg/kg; whereas SM showed obvious toxicity symptoms. This research demonstrated the potential of industrial hemp to remediate Pb-contaminated mining soils.

Wang et al. (Contribution 6) evaluated the combined effects of Cd and F on lettuce growth and soil bacterial communities in co-contaminated yellow soil. Low concentrations of Cd and F had no significant effect on lettuce growth, but synergistic negative effects were observed when F exceeded 300 mg/kg and Cd exceeded 1.0 mg/kg. Lettuce accumulated Cd and F mainly in roots, with translocation factors (TrF) ranging from 0.313–0.941 to 0.499–0.754, respectively. This study highlighted the importance of considering combined pollution in remediation practices.

2.3. Soil Amendment and Immobilization Technologies

Two studies focused on soil amendment technologies, exploring the potential of waste-derived materials to immobilize heavy metal.

Cecire et al. (Contribution 7) investigated the use of rice husk as a sustainable amendment for heavy metal immobilization. Rice husk showed high retention capacities for

https://doi.org/10.3390/toxics14020154

Cu (100%), Cd (64%), and Mn (18%) under optimal conditions (particle size 90–300 μm, pH > 5.5, low buffer concentration). Pot experiments demonstrated that rice husk reduced heavy metal uptake by Lactuca sativa and Spinacia oleracea by 40–60% for Mn and Zn, and nearly 100% for Cr, Cu, Ni, Cd, and Pb. Factors such as abundance, low cost, and environmental friendliness make the material a promising option for soil remediation.

Rashid et al. (Contribution 8) developed machine learning models to predict Cd transformation and immobilization in biochar-amended soils. Three models (LSTM, Bi-GRU, and 5-layer CNN) were tested, with the 5-layer CNN achieving the highest accuracy (R^2 = 0.956) in biochar-amended soils. Key factors that were identified as influencing Cd immobilization were as follows: soil pH, organic carbon (OC), cation exchange capacity (CEC), electrical conductivity (EC), and clay content. Biochar effectively reduced Cd availability by altering soil properties, providing a data-driven approach for the optimization of biochar application.

2.4. *Microbial Responses and Ecological Restoration*

One article focused on the response of soil microbial communities to heavy metal stress, shedding light on the ecological mechanisms of soil restoration.

Wang et al. (Contribution 6) studied the effect of Cd, F, and their combination on soil bacterial communities in lettuce rhizosphere. Proteobacteria was the dominant phylum (33.42–44.10%), and compartment (rhizosphere vs. bulk soil) was the primary factor driving community variation, followed by pollutant stress. F and Cd showed synergistic effects on bacterial communities, with the rhizosphere enriching specific taxa (e.g., Oxyphotobacteria, Subgroup 6) to enhance plant resistance. This study emphasized the role of soil microorganisms in mediating plant–pollutant interactions during remediation.

3. Key Insights and Future Directions

This Special Issue showcases the diversity and depth of research in heavy metal polluted farmland management and offers several key insights:

First, accurate source identification is the premise of effective remediation. The combination of multivariate statistical methods (PCA, PMF) and field surveys proves effective in distinguishing natural and anthropogenic sources, providing targeted guidance for pollution control. Second, phytoremediation demonstrates great potential when selecting appropriate plant species and optimizing cultivation strategies. Chrysanthemums and industrial hemp demonstrate both high remediation efficiency and economic benefits, addressing the long-standing problem of low profitability in phytoremediation. Third, soil amendments derived from agricultural waste (e.g., rice husk, biochar) offer sustainable and cost-effective options for heavy metal immobilization, aligning with the principles of circular economy. Fourth, the response of soil microbial communities to heavy metal stress plays a crucial role in ecological restoration; manipulation of rhizosphere microorganisms could further enhance remediation efficiency.

Despite these advances, several challenges and future directions remain:

1. Combined pollution remediation: Most studies focused on single or a few heavy metals, despite the fact that farmland is often contaminated by multiple PTEs. Future research should explore the mechanisms of combined pollution and develop synergistic remediation technologies [6].
2. Long-term field validation: Many studies are conducted under controlled conditions; long-term field trials are needed to verify the stability and durability of remediation effects under natural environmental fluctuations [7].

https://doi.org/10.3390/toxics14020154

3. Integration of multiple technologies: The combination of phytoremediation, microbial remediation, and soil amendment could improve remediation efficiency. For example, the co-application of biochar and hyperaccumulators shows promising results [8].
4. Risk assessment and safe utilization: More research is needed to establish thresholds for the safe utilization of polluted farmland, integrating crop quality, soil health, and human health risks [9].
5. Digitalization and intelligent management: Machine learning models and digital tools should be further developed to optimize remediation strategies, predict pollution trends, and promote the traceability of agricultural products from polluted areas [10].

4. Conclusions

This Special Issue provides a comprehensive overview of the latest research progress in the safe utilization and ecological restoration of heavy metal polluted farmland. The collected studies cover contamination characterization, source identification, phytoremediation, soil amendment, microbial responses, and digital modeling, offering practical solutions and theoretical support for addressing this global challenge.

Notably, all studies are based on case studies from China, a country facing severe farmland heavy metal pollution due to rapid industrialization and intensive agriculture. The research findings not only contribute to solving local environmental problems but also provide valuable references for other regions with similar challenges. This Special Issue highlights the importance of an all-encompassing approach to the management of heavy metal polluted farmland, by integrating ecological principles, economic benefits, and technological innovations.

As we move forward, continuous research and innovation are essential to develop more efficient, sustainable, and cost-effective remediation technologies. It is also crucial to strengthen policy support, promote interdisciplinary collaboration, and enhance public awareness to achieve the dual goals of environmental protection and food security. We hope this Special Issue will inspire more researchers and practitioners to contribute to the safe utilization and ecological restoration of heavy metal polluted farmland, advancing towards a more sustainable and resilient agricultural system.

Author Contributions: Writing—original draft preparation, B.G.; writing—review and editing, Y.F. All authors have read and agreed to the published version of the manuscript.

Data Availability Statement: Not applicable.

Acknowledgments: We would like to express our sincere gratitude to all authors for their valuable contributions to this Special Issue. We also thank the reviewers for their rigorous and constructive comments, which greatly improved the quality of the published articles. Special thanks to the editorial team of Toxics for their professional support throughout the publication process. This work was supported by various funding agencies mentioned in the individual articles, whose contributions are greatly appreciated.

Conflicts of Interest: The authors declare no conflicts of interest.

List of Contributions:

1. Cao, M.; Jia, Y.; Lu, X.; Huang, J.; Yao, Y.; Hong, L.; Zhu, W.; Wang, W.; Zhu, F.; Hong, C. Temporal and Spatial Variation of Toxic Metal Concentrations in Cultivated Soil in Jiaxing, Zhejiang Province, China: Characteristics and Mechanisms. *Toxics* **2024**, *12*, 390.
2. Chen, S.; Wang, H.; Han, R. Source Apportionment of Potentially Toxic Elements in Agricultural Soils of Yingtan City, Jiangxi Province, China: A Principal Component Analysis–Positive Matrix Factorization Method. *Toxics* **2025**, *13*, 267.

3. Wang, M.; Yu, P.; Tong, Z.; Shao, X.; Peng, J.; Hamid, Y.; Huang, Y. A Modified Model for Quantitative Heavy Metal Source Apportionment and Pollution Pathway Identification. *Toxics* **2024**, *12*, 382.
4. Lu, X.; Chen, Y.; Song, J.; Bao, J.; Dai, C.; Sun, R.; Liu, J.; Jin, C.; Zhong, N.; Huang, C.; et al. Screening of Profitable Chrysanthemums for the Phytoremediation of Cadmium-Contaminated Soils. *Toxics* **2025**, *13*, 360.
5. Xu, Y.; Kumpeangkeaw, A.; An, X.; Chen, X.; Zhang, Y.; Lv, P.; Zhang, Q.; Guo, R.; Ji, Q.; Yang, M. The Tolerance Differences of Two Industrial Hemp Varieties Under Lead (Pb) Stress. *Toxics* **2025**, *13*, 90.
6. Wang, M.; Chen, X.; Hamid, Y.; Yang, X. Evaluating the Response of the Soil Bacterial Community and Lettuce Growth in a Fluorine and Cadmium Co-Contaminated Yellow Soil. *Toxics* **2024**, *12*, 459.
7. Cecire, R.; Diana, A.; Giacomino, A.; Abollino, O.; Inaudi, P.; Favilli, L.; Bertinetti, S.; Cavalera, S.; Celi, L.; Malandrino, M. Rice Husk as a Sustainable Amendment for Heavy Metal Immobilization in Contaminated Soils: A Pathway to Environmental Remediation. *Toxics* **2024**, *12*, 790.
8. Rashid, M.S.; Wang, Y.; Yin, Y.; Yousaf, B.; Jiang, S.; Mirza, A.F.; Chen, B.; Li, X.; Liu, Z. Quantitative Soil Characterization for Biochar–Cd Adsorption: Machine Learning Prediction Models for Cd Transformation and Immobilization. *Toxics* **2024**, *12*, 535.

References

1. Hou, D.; Zhang, Y.; Li, J.; Wang, S.; Smith, A.; Liu, C.; Olsen, S.; Zhao, F. Global Soil Pollution by Toxic Metals Threatens Agriculture and Human Health. *Science* **2025**, *388*, 1452–1456. [CrossRef] [PubMed]
2. Li, Y.; Tao, H.; Cao, H.; Wan, X.; Liao, X. Achieving Synergistic Benefits Through Integrated Governance of Cultivated Cadmium Contamination via Multistakeholder Collaboration. *Nat. Commun.* **2024**, *15*, 9817. [CrossRef] [PubMed]
3. Li, H.X.; Cui, X.L.; Sun, Y.C.; Zheng, P.; Wang, L.; Shi, X.Y. Advances in Microbial Remediation of Heavy Metal-Contaminated Soils: Mechanisms, Synergistic Technologies, Field Applications and Future Perspectives. *Toxics* **2025**, *13*, 1069. [CrossRef] [PubMed]
4. Xu, L.; Zhao, F.F.; Xing, X.Y.; Peng, J.B.; Wang, J.M.; Ji, M.F.; Li, B.L. A Review on Remediation Technology and the Remediation Evaluation of Heavy Metal-Contaminated Soils. *Toxics* **2024**, *12*, 897. [CrossRef] [PubMed]
5. Yang, L.; Wang, J.; Yang, Y.; Li, S.; Wang, T.; Oleksak, P.; Chrienova, Z.; Wu, Q.; Nepovimova, E.; Zhang, X.; et al. Phytoremediation of heavy metal pollution: Hotspots and future prospects. *Ecotoxicol. Environ. Saf.* **2022**, *234*, 113403. [CrossRef] [PubMed]
6. Xiong, X.; Wang, J.; Liu, J.; Xiao, T. Microplastics and Potentially Toxic Elements: A Review of Interactions, Fate and Bioavailability in the Environment. *Environ. Pollut.* **2024**, *340*, 122754. [CrossRef] [PubMed]
7. Wan, X.; Lei, M.; Yang, J.; Chen, T. Three-Year Field Experiment on the Risk Reduction, Environmental Merit, and Cost Assessment of Four In Situ Remediation Technologies for Metal(loid)-Contaminated Agricultural Soil. *Environ. Pollut.* **2020**, *266*, 115193. [CrossRef] [PubMed]
8. Li, K.; Yang, B.; Wang, H.; Xu, X.; Gao, Y.; Zhu, Y. Dual Effects of Biochar and Hyperaccumulator *Solanum nigrum* L. on the Remediation of Cd-Contaminated Soil. *PeerJ* **2019**, *7*, e6631. [CrossRef] [PubMed]
9. Peng, J.Y.; Zhang, S.; Han, Y.; Bate, B.; Ke, H.; Chen, Y. Soil Heavy Metal Pollution of Industrial Legacies in China and Health Risk Assessment. *Sci. Total Environ.* **2022**, *816*, 151632. [CrossRef] [PubMed]
10. Duan, Z.; Chen, C.; Ni, C.; Xiong, J.; Wang, Z.; Cai, J.; Tan, W. How Different Is the Remediation Effect of Biochar for Cadmium Contaminated Soil in Various Cropping Systems? A Global Meta-Analysis. *J. Hazard. Mater.* **2023**, *448*, 130939. [CrossRef] [PubMed]

https://doi.org/10.3390/toxics14020154

toxics

MDPI

Article

A Modified Model for Quantitative Heavy Metal Source Apportionment and Pollution Pathway Identification

Maodi Wang [1], Pengyue Yu [1], Zhenglong Tong [1], Xingyuan Shao [1], Jianwei Peng [1], Yasir Hamid [2] and Ying Huang [1,*]

[1] National Engineering Laboratory of High Efficient Use on Soil and Fertilizer, College of Resources, Hunan Agricultural University, Changsha 410128, China; w1002043947@163.com (M.W.); 13203130035@163.com (P.Y.); 13203618886@stu.hunau.edu.cn (Z.T.); slock130@stu.hunau.edu.cn (X.S.); jianweipenglab@hunau.edu.cn (J.P.)

[2] Ministry of Education (MOE) Key Lab of Environment, Remediation and Ecological Health, College of Environmental and Resources Science, Zhejiang University, Hangzhou 310058, China; yasirses2007@gmail.com

* Correspondence: huangying@hunau.edu.cn; Tel.: +86-0731-84617803

Abstract: Current source apportionment models have successfully identified emission sources and quantified their contributions. However, when being utilized for heavy metal source apportion in soil, their accuracy needs to be improved, regarding migration patterns. Therefore, this work intended to improve the pre-existing principal component analysis and multiple linear regression with distance (PCA-MLRD) model to effectively locate pollution pathways (traffic emissions, irrigation water, atmospheric depositions, etc.) and achieve a more precise quantification. The dataset of soil heavy metals was collected from a typical area in the Chang-Zhu-Tan region, Hunan, China in 2021. The identification of the contribution of soil parent material was accomplished through enrichment factors and crustal reference elements. Meanwhile, the anthropogenic emission was identified with principal component analysis and GeoDetector. GeoDetector was used to accurately point to the pollution source from a spatial differentiation perspective. Subsequently, the pollution pathways linked to the identified sources were determined. Non-metal manufacturing factories were found to be significant anthropogenic sources of local soil contamination, mainly through rivers and atmospheric deposition. Furthermore, the influence of irrigation water on heavy metals showed a more pronounced effect within a distance of 1000 m, became weaker after that, and then gradually disappeared. This model may offer improved technical guidance for practical production and the management of soil heavy metal contamination.

Keywords: heavy metals; source apportionment; receptor model; grouped principal components; GeoDetector

1. Introduction

The rapid industrial and economic developments worldwide have led to significant challenges concerning heavy metals contamination [1,2]. Nearly 20% of farmlands in China exhibit varying degrees of contamination by Cd and Pb [3,4]. The first step to address the soil heavy metals pollution is the identification and regulation of contaminants sources. Currently, soil heavy metal pollution source analysis can be categorized into qualitative and quantitative approaches. Among the quantitative analyses, principal component analysis and multiple linear regression (PCA-MLR), positive principal matrix decomposition (PMF), and the absolute principal component score are commonly employed [5]. In our recent study, we introduced a distance variable to the receptor model and proposed a modified receptor model based on principal component analysis (PCA) and multiple linear regression with distance (MLRD) [6]. The model follows the fact that the spread of heavy metals from emission sources to soil diminishes as the distance increases, and the extent of this contribution can be quantified with the distance between sampling sites and the sources.

This novel model efficiently identifies the precise pollution sources and quantifies their respective contributions. Previously, a number of studies have employed this model. For instance, Zeng et al. [7] used PCA-MLRD to conduct a source analysis of soil heavy metals near a multifunctional industrial park in Anhui Province, China. This approach successfully identified the possible source locations and source types, thereby reducing the subjectivity of potential source screening. However, this model still has two unsolved issues; firstly, it failed to account for the vertical migration of heavy metals, and secondly, it could not identify pollution processes related to each source.

In addressing the issue of source identification, it is worth noting that principal component analysis has been commonly employed in previous studies. However, it is important to note that the resulting explanatory elements of the factors tend to exhibit a certain degree of similarity. To cope with this issue, researchers have proposed the concept of grouped principal component analysis (GPCA), which extends the application of PCA and enables the categorization of components into distinct groups of similar characteristics, thus allowing for more precise observations [8]. Regarding matching of emission sources to obtained factors, previous researchers used typical indicators for different types of emissions (for example, Pb for traffic emissions, Cd for industrial activities, and so on), which may lead to inaccurate results due to the overly subjective judgment. Furthermore, various factories may display variability in the release of different kinds of heavy metals. For example, the production of clothing involves the utilization of raw materials with high levels of Ni and Cr [9]. Soils near chlor-alkali factories carry high levels of Hg [10], while lead smelters are associated with high levels of Pb contamination in top soils [11]. Therefore, it is necessary to make accurate judgments of potential emissions instead of identifying natural or industrial source types.

Heavy metals from pollution sources are released through sewage discharge, solid waste, and exhaust gases, which can pollute the soil through water irrigation and atmospheric deposition, as well as solid waste diffusion [12–15]. Atmospheric deposition is a global problem with serious implications for the environment and human health [16–18]. Previous studies have emphasized the significance of atmospheric deposition as a major contributor to heavy metals accumulation in soils [19,20]. It is well established that a reduction in atmospheric heavy metals input into agricultural soils can effectively decrease the concentration of metals in crops [21]. Researchers analyzing the sources of heavy metals in French farmland soils found that atmospheric deposition is the main input for Cu, Ni, Pb, and Zn accumulation in soils [22]. Atmospheric deposition is the main source for heavy metals accumulation in farmland in England and Wales, accounting for between 25% and 85% of the total input [23]. In recent years, developing countries have gradually begun to pay attention to the pollution problems caused by atmospheric deposition. In China, industrialization has made atmospheric deposition an intermediate pathway for the migration of heavy metals, leading to more severe heavy metal pollution in the soil [24]. It has been reported that atmospheric deposition was an important source for heavy metals input in agricultural soils in Hunan Province, which accounted for 51.24–94.74% of the total input [25,26]. However, a limited number of studies have included atmospheric depositions into receptor models. Therefore, it is necessary to determine the pollution pathways of sources contributing to heavy metals accumulation in soils, and to achieve more accurate source apportionment.

To solve the aforementioned issues, this research introduced modifications to the PCA-MLRD model through utilizing the GPCA method and GeoDetector. The study was designed to accomplish the following three tasks: (1) identification of sources and quantification of the emission contributions, (2) identifying the pollution pathways while quantifying the impacts, (3) comparing the accuracy with previously existing models.

2. Materials and Methods

2.1. Overview of the Research Area

The study area consists of several townships belonging to 6 connected cities in Hunan Province, China, with a total area of 1333 km^2, located at 27°24′–28°50′ N and 111°10′–114°15′ E. The study region features a subtropical monsoon climate characterized by prevailing southeast winds during summer and northeast winds in winter. The average annual temperature range is 16 °C–28 °C, with an average annual humidity of 70~80%, and an average annual precipitation estimate ranging from 1400 to about 1700 mm. The distribution maps of lithology, soil type, land use, and water are listed in Figure 1 (http://www.geodata.cn; accessed on 27 September 2022). The sampling sites were all located in paddy fields, which were dependent on the third- and fourth-grade rivers for irrigation. The study area contains six types of soils, yellow earths, latosols, weakly developed red earths, red earths, yellow red earths, and paddy soils.

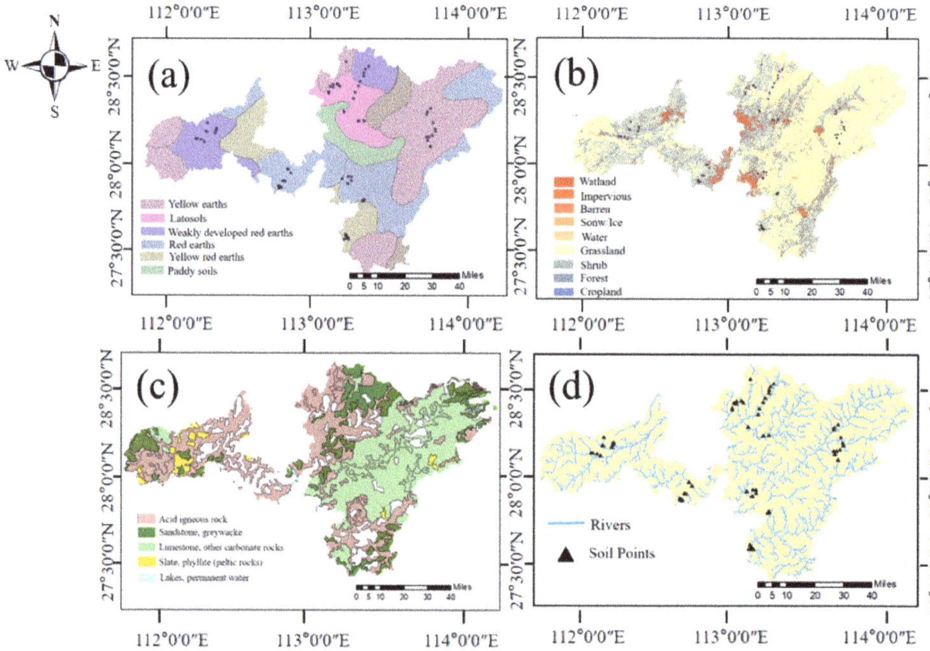

Figure 1. Spatial distribution map of soil types (**a**), land use (**b**), lithology (**c**), and water systems (**d**).

2.2. Sampling and Analysis

In this study, a total of 109 soil samples were collected from the cultivated layer (0–15 cm) of paddy fields. Each sampling location was recorded using a portable global positioning system (GPS), and the spatial distribution of the sampling points is illustrated in Figure 2. At the same time, a set of forty-nine dustfall cylinders (height: 30 cm; diameter: 16 cm) were placed on the roofs of residential buildings near the sampling sites to capture atmospheric depositions. The atmospheric deposition samples were collected four times throughout the year, once in each of the four seasons. During the sample collection process, the dust adhering to the bottom and inner wall of the cylinder was carefully brushed using a nylon test tube brush and thoroughly mixed with the liquid sample present in the cylinder. The resulting samples were collected in PET plastic bottles (1 L) following the addition of 10 mL of highly pure HNO_3 and subsequently transported to the laboratory for analysis.

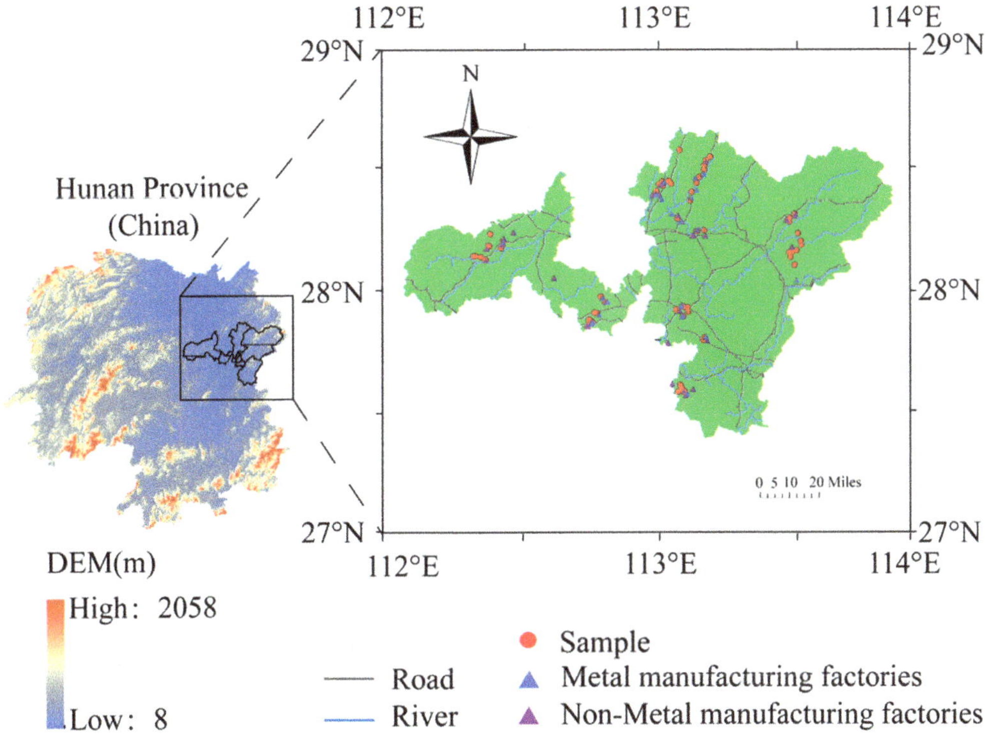

Figure 2. The map of the study area and locations of sampling points.

All soil samples were dried naturally, and the larger plants and animal residues were removed from the soil samples using plastic tweezers. Then, the soil samples were ground and passed through a 100-mesh polyethylene sieve (0.15 mm). About 0.200 g of soil was weighed and subjected to digestion on a hot plate by mixing with HNO_3-H_2O_2-HF (6:3:3). Analytically pure (AR)-grade reagents were used for digestion and were obtained from Sinopharm Chemical Reagent Co., Ltd. (Shanghai, China) The atmospheric deposition samples were separated into dry and aqueous ones through filtration. Dry samples were air-dried and digested in the same way as the soil samples. The aqueous samples were extracted at 50 mL and acidified by hydrochloric acid, then digested with 2 mL HNO_3 and 1 mL HCl at 85 °C until the samples evaporated to 20 mL. The concentrations of different elements including Cd, Ni, Cu, Zn, Cr, As, Pb, Mg, and Ca in the digested solution were detected using inductively coupled plasma mass spectrometry (ICP-MS, model NEXION 350 X; PerkinElmer, Waltham, MA, USA). To ensure the accuracy of the chemical analysis, three replicate tests, and blank reagents, as well as standard reference materials (GBW07429, China National Reference Material Research Center for Reference Materials, Beijing, China) were set up for the determination, and the recoveries of all analyses fell within the range between 90 and 110%.

2.3. *Model Hypothesis*

This study assumed that heavy metals in soil come from the parent materials and anthropogenic emissions. The anthropogenic emissions can be divided into point sources, such as factories, and non-point sources, such as traffic emissions, irrigation water, and atmospheric deposition. The main types of factories include chemical factories (cement factories, fireworks factories) that mainly emit Cd, Pb [27–29]; metal manufacturing factories (steel factories, electroplating factories) mainly emit Cu, As, Zn [30,31]; non-metal manufacturing factories (textile factories) mainly emit As, Cr, and Ni [32,33]. The contributions of parent materials were calculated with the enrichment factor and crustal reference element. All point sources (factories) in this area could directly emit heavy metals as sewage discharge, solid waste, and exhaust gases. In the meantime, all rivers and roads could provide channels to transfer heavy metal from the source to the sink (soil) as irrigation water and traffic emissions, where the influence becomes weaker with the increase in distance. Therefore, in the data collection, the distribution of all factories and rivers and roads was investigated. The distance refers to the shortest Euclidean distance from sampling site to factory, river, or road. In detail, traffic emissions and irrigation water are not regarded as emission sources; with the ban on leaded petrol, heavy metals from traffic emissions come mainly from the transport of industrial products and solid waste [34,35]. Therefore, to better reflect the impact of roads, traffic emissions are used as the pollutant pathway. Heavy metals in irrigation water come from industrial emissions and are, therefore, also set as a pollution pathway [36]. It has been shown that the vast majority of heavy metals in the atmosphere originate from industrial activities; they then enter the soil by means of atmospheric deposition [15,37]. Atmospheric deposition is, therefore, also studied as a pathway of industrial pollution, rather than as a source of pollution. Overall, the model includes soil-forming parent material sources and anthropogenic sources, including point and non-point sources, in its calculations.

2.4. Quantification of the Contribution of Soil Parent Material

Differences in lithology and soil parent material are a major cause of spatial heterogeneity of heavy metal in soils. Meanwhile, soil parent material is also thought to be the most important source of heavy metals. Therefore, the lithology of the study area was analyzed, and the contribution of soil parent material was firstly quantified with enrichment factors and crustal reference elements as the following:

$$\mathrm{EF} = \frac{\left(\frac{[\mathrm{M}]}{[\mathrm{C}]}\right)_{\mathrm{Soil}}}{\left(\frac{[\mathrm{M}]}{[\mathrm{C}]}\right)_{\mathrm{Crust}}} \tag{1}$$

$$[\mathrm{M}]_{\mathrm{excessive}} = [\mathrm{M}]_{(\mathrm{soil})} - [\mathrm{C}]_{(\mathrm{Soil})}\left(\frac{[\mathrm{M}]}{\mathrm{C}}\right)_{\mathrm{Crust}} \tag{2}$$

where EF: enrichment factors; [M]: the content of trace element; [C]: content of crustal reference element; $\left(\frac{[\mathrm{M}]}{[\mathrm{C}]}\right)_{\mathrm{Soil}}$: the ratio of concentrations of trace elements in collected soil samples to crustal reference elements; $\left(\frac{[\mathrm{M}]}{[\mathrm{C}]}\right)_{\mathrm{Crust}}$: the ratio of background values of each trace element to crustal reference elements.

2.5. Identification of Anthropogenic Sources and Their Potential Pathways with GPCA and GeoDetector

The application of principal component analysis (PCA) for grouping purposes sometimes encounters difficulties in identifying complex factors, particularly when dealing with a large number of elemental species, so the repetition of factor-loading elements can lead to inaccurate judgments. Therefore, in order to achieve a comprehensive division of mixed sources, the grouped principal component analysis (GPCA) method is preferred over the principal component analysis (PCA) method.

GeoDetector is a novel statistical method utilized to identify spatial heterogeneity and unveil the driving factors behind it [38–40]. Therefore, in this study, GeoDetector was utilized as a tool to find out the main pollution pathways of each anthropogenic source. The GeoDetector model examines the relationships between dependent variables such as heavy metal contents in soil and the explanatory variables, which include atmospheric deposition, altitude, distance to factories, and distance to roads and rivers. The altitude data corresponds to the elevation at each sampling point and was collected from the geospatial data cloud platform (http://www.gscloud.cn/; accessed on 27 September 2022). On the other hand, the distance factor represents the minimum Euclidean distance to the pollution source as well as nearby roads and rivers. It is worth noting that the explanatory variables used in GeoDetector are categorical variables. Specifically, the distance to the road, the distance to the river, the distance to the factory were discretized as classification variables, and the continuous variables should be discretized. The distances were categorized based on equal intervals of 200 m, while the elevation and atmospheric deposition were discretized into 10 categories by natural breaks. Based on the above explanatory variables and the dependent variable Y for factor detection and interaction detection, the factor detector uses the q value to measure the extent of the impact that factor (X) has on the dependent variable Y. The q values fall within the range [0, 1], with a larger q value indicating stronger explanatory power [41], and its calculation process is presented in Equation (3).

$$q = 1 - \sum_{h=1}^{L} N_{R_h} \frac{\sigma_{R_h}^2}{N_R \sigma_R^2} \tag{3}$$

where N_{R_h} and N_R represent the grid numbers of the R_h partition and the entire study area R, respectively. $\sigma_{R_h}^2$ and σ_R^2 are the variances of elemental heavy metal accumulation within the R_h partition and the complete study area R, respectively.

For the interaction detector, when $q(X1 \cap X2) > q(X1) + q(X2)$ for two explanatory variables, it indicates that the two factors exhibit a nonlinear connection with each other. In this study, the pollution path of the source is determined by the two explanatory variables with higher q values of nonlinear enhancement following interaction with different types of factories.

2.6. Quantification of the Contributions from Anthropogenic Emissions

On the basis of the PCA-MLRD model [6], the influence of pollution pathways was added to the modified model. And the modification will enhance the simulation of the pollution processes associated with emission sources. The calculation process is shown in Equation (4):

$$[M]_{excessive} = \sum_{j=1}^{p} B_{in} P_{nm} D_{mk} \tag{4}$$

where $[M]_{excessive}$ represents the cumulative content of heavy metals; D_{mk} represents the distance from the source to the sampling point ; P_{nm} represents the main pollution pathways of source n; B_{in} is the regression coefficient; n is the number of pollution sources. The mathematical transformations used in this study include logarithmic and exponential transformations for the distances matrix to find the best form of fit. In order to avoid collinearity, the stepwise regression method is used to obtain fitting parameters in multiple linear regression.

In order to maintain the accuracy of the model, samples (109) were divided into two sets: fitting sets and test sets, with the proportion of 80% (87) and 20% (22), respectively. It should be noted that the selection of fitting set and test set data was not done manually, but was carried out by using the random selection function in Excel. Moreover, four team members individually conducted the modeling separately from the start of the dataset, including the selection of crustal reference elements and model fitting. This process helps greatly reduce the subjectivity brought by artificial screening and improves the credibility of the model. Moreover, $R^2 > 0.5$ was set as a basis for the test in the test set data.

2.7. Data Analysis

The pre-processing and statistical analysis were carried out using Microsoft Excel 2003. Graphs were created in Origin 2019 and ArcGIS 10.6. Principal component analysis and multiple linear regression for soil heavy metals were computed using SPSS 19.0 software, with the principle of eigenvalues > 1 (Kaiser criterion) to filter the principal components, while factor loadings were determined with varimax rotation [42]. ArcGIS 10.6 was used for spatial distribution mapping and shortest Euclidean distance calculation, and the inverse distance weighting (IDW) interpolation method was used for spatial analysis.

3. Results

3.1. The Statistical Analysis of Soil Heavy Metals and Quantification of Natural Sources

Descriptive statistics of soil are displayed as the maximum concentration of all the heavy metals that exceeded the standard values, with notably elevated levels of Cd and As, which exceed the standard values by 8.3 and 2.1 times, respectively (Table 1). The average concentration of heavy metals including As, Cd, Cr, Cu, Ni, Pb, Zn, Ca, and Mg in the study area were observed to be 26.47, 0.70, 87.07, 29.37, 28.41, 43.94, 133.78, 1471.28, and 4600.78 mg/kg, respectively. It was found that the mean Cd concentration exceeded the standard by 1.8 times. The ranking of heavy metals according to their coefficients of variation (CV) followed the order of Cd (57.21%), As (51.27%), Cr (42.05%), Pb (35.57%), Cu (33.99%), Ni (31.59%), and Zn (26.83%). The elemental exceedance rates were highest for Cd at 90% and smallest for Ni at 2%.

Table 1. Descriptive statistics of heavy metals contents in soil of the study area (mg/kg).

Element	As	Cd	Cr	Cu	Ni	Pb	Zn	Ca	Mg
Minimum (mg/kg)	5.85	0.20	16.33	13.33	10.78	17.85	73.48	642.73	2662.55
Median (mg/kg)	26.72	0.54	81.89	27.42	27.8	40.35	131.97	1254.23	4444.38
Maximum (mg/kg)	83.81	2.48	219.75	65.46	71.95	95.85	305.63	5966.32	10,447.30
Mean (mg/kg)	26.47	0.70	87.07	29.37	28.41	43.94	133.78	1471.28	4600.78
Skewness	1.04	1.38	1.07	1.34	1.47	0.93	1.13	2.84	1.53
Kurtosis	2.21	2.69	1.42	2.57	5.19	0.84	3.86	10.84	3.85
SD [a]	13.57	0.40	36.61	9.98	8.98	15.63	35.9	851.34	1323.58
CV% [b]	51.27	57.21	42.05	33.99	31.59	35.57	26.83	57.86	28.77
Risk screening (mg/kg)	40.00	0.30	150.00	50.00	60.00	70.00	200.00	-	-
Background [c] (mg/kg)	12.80	0.14	150.00	25.00	27.80	30.00	84.20	1300.00	4000.00
Excess rate [d]	16%	90%	8%	5%	2%	9%	5%	-	-

Note: [a] SD means standard deviation; [b] CV means coefficient of variation; [c] background values of heavy metals in Hunan Province; [d] percentage of points with soil heavy metal contents exceeding background values in Hunan Province.

3.2. Analysis of Potential Pollution Sources by GPCA and Distributions of Factories

PCA analysis identified three principal components, which collectively accounted for 63.85% of the total variance. Elemental loads greater than 0.7 are considered major load elements. The main interpretive elements of PCA1 include As, Cd, Pb, while those of PCA2 are Ni and Cr, and that of PCA3 is Mg (Figure 3a). However, out of the three factors mentioned above, PCA1 had four loading elements that were repeated, and the remaining two factors did not show repeated elements. Moreover, the GPCA method was used to further decompose the PCA1. The main explanatory elements of PCA1-1 are Cd (0.83), Pb (0.90), while those of PCA1-2 include As and Zn (Figure 3b). Therefore, by combining both PCA and GPCA methods, a total of four sources were identified, namely PCA1-1, PCA1-2, PCA2, PCA3.

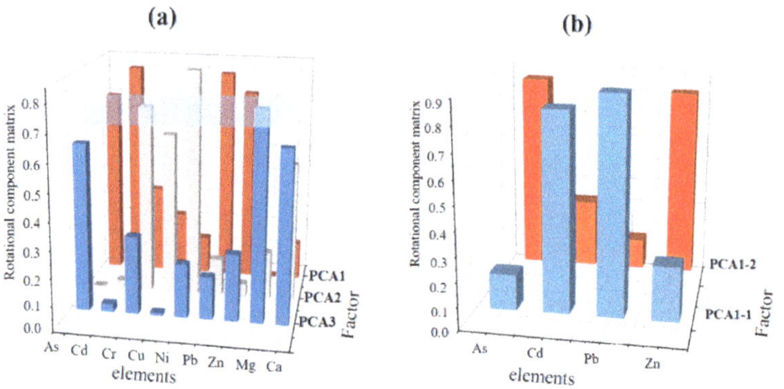

Figure 3. Rotated component matrices obtained by PCA (**a**) and GPCA (**b**).

Figure 4 shows the spatially interpolated maps of factor scores derived from the GPCA model and spatial distribution maps of different types of factories in the studied region. The study area is predominantly characterized by the presence of cement manufacturers and fireworks factories, which were densely dispersed in the eastern and central parts. Additionally, the factor scores of PCA1-1 exhibited higher scores in the central and eastern parts of the spatial interpolation map. Meanwhile, metal manufacturing factories were predominant in the northern and central regions of the study area, aligning with higher PCA1-2 factor scores. In addition, locations with relatively high factor scores for PCA2-1 were found to have non-metal manufacturing factories, including textile factories. In general, there exists a positive relevance between higher factor scores with a denser distribution of factories and regions with large factories. The comparison revealed that the spatial interpolation of PCA1-1, PCA1-2, and PCA2 factor scores had similar patterns to the spatial distribution of chemical factories, non-metal manufacturing factories, and metal manufacturing factories in the study area.

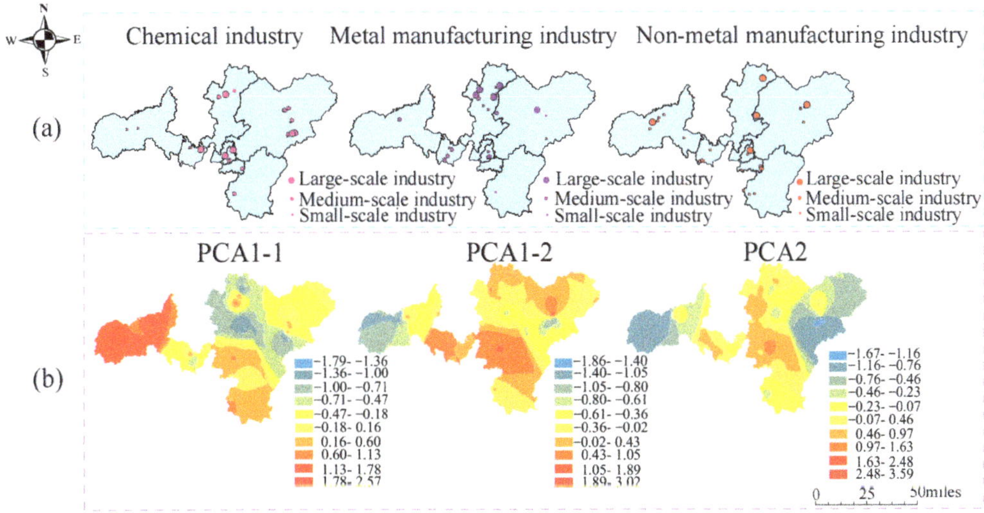

Figure 4. Comparison of the spatial distribution of three different types of factories (**a**) with the spatial interpolation of factor scores of the decomposed PCA model (**b**).

3.3. Analysis of Pollution Source Pathways Based on GeoDetector

As shown in Figure 5, the distance to chemical plants in PCA1-1 has the highest degree of explanation (0.43); the distance to metal manufacturing plants in PCA1-2 has the highest degree of explanation (0.56) and is significantly higher than the rest of the variables; the distance to non-metal manufacturing plants in PCA2 has the highest degree of explanation (0.47); the soil type in PCA3 has the highest degree of explanation (0.47), and the land use type also has a some degree of explanation (0.26).

In order to identify potential contamination pathways, a combination of the factor detection and interaction detection modules of GeoDetector was employed. The identification of influencing factors was executed through GeoDetector (Figure 6a), while Figure 6b portrays the results of multiple-factor interaction detections. A large number of q values after the interaction confirmed the increment compared to the factor detection results in Figure 5. For Cd, the distance to non-metal manufacturing factories had a significant interaction with atmospheric deposition and roads, as evidenced by q values of 0.52 and 0.44, respectively. These findings suggest that Cd emissions from non-metal manufacturing factories facilitated the soil's contamination by two pollution pathways, including atmospheric deposition and irrigation water. Similarly, the interactions between the distance of the chemical factories and the roads' distance and atmospheric deposition indicated that traffic emissions and atmospheric deposition are the main pollution pathways for Cd emissions from chemical factories. Meanwhile, Cd pollution caused by emissions from metal manufacturing factories was affected by atmospheric deposition. In addition, the same analytical approach was employed to assess the major pollution paths of Pb and As emissions from different factories. For example, Pb pollution from non-metal manufacturing factories was influenced by atmospheric deposition, while chemical and metal manufacturing factories were primarily influenced by traffic emission and irrigation water. Concerning As pollution, non-metal manufacturing factories were linked to irrigation water and atmospheric deposition, chemical factories were associated with traffic emission, and metal manufacturing factories demonstrated a connection with traffic emission and irrigation water.

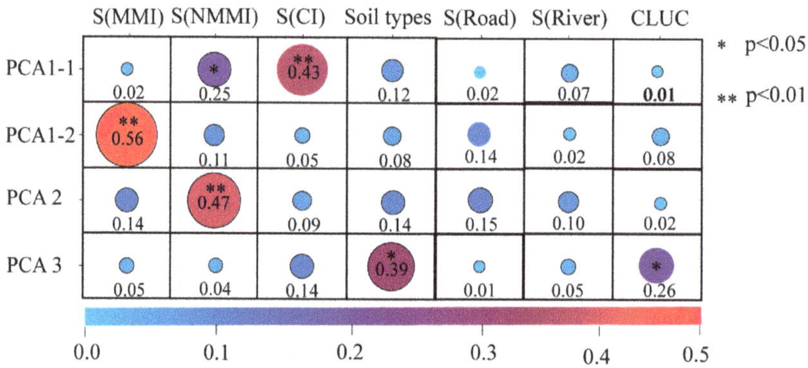

Figure 5. Identification results (q values) of principal component factor types using GeoDetector. Note: NMMI: non-metal manufacturing factory; CI: chemical factory; MMI: metal manufacturing factory; CLUC: classification of land use.

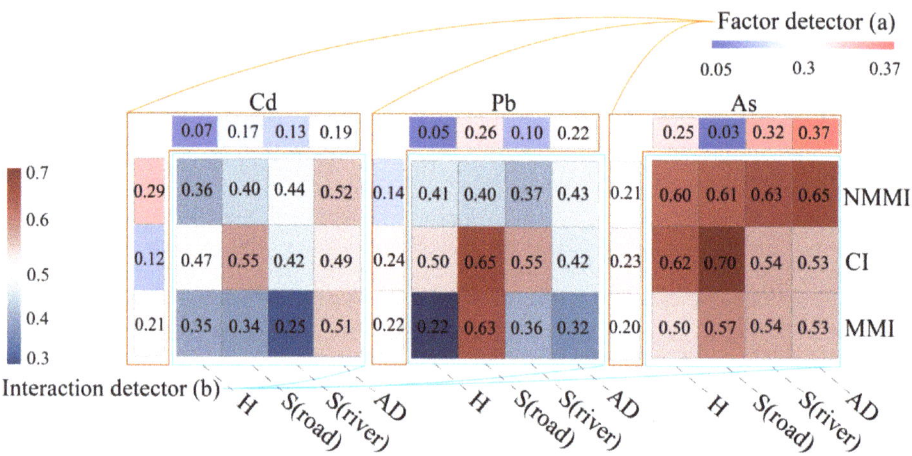

Figure 6. Analysis of paths by factor detectors (**a**) and interaction detectors (**b**). Note: NMMI: non-metal manufacturing factory; CI: chemical factory; MMI: metal manufacturing factory; H: altitude; AD: atmospheric deposition.

3.4. The Quantification of Contribution from Anthropogenic Sources

As shown in Table 2, four models were conducted separately by four team members. In detail, Mg was used as a crustal element, and the national background values were used in model 1; Zn and Mg were used as crustal elements and the background values for different soil types were used in model 2 and model 3 (Background values for different soil types are shown in Table S1), respectively. Model 4 is the model obtained by fitting using PCA-MLRD. The conversion of the distance matrix and regression processes were both carried out separately. Equations (5)–(7) represent the regression of model 3 for Cd, As, and Pb. It can be seen that, despite the variation in contribution, the identification of sources was nearly the same. The values of R^2 and RMSE all showed a better fitting of modified models than the original PCA-MLRD (Table 2). Based on the fitting results, it can be seen that the three anthropogenic pollution sources were finally found by the modified PCA-MLRD model, and together with the contribution of the background values, the contribution of each source was calculated separately (Figure 7). The Cd pollution was primarily caused by metal manufacturing factories (30–44%), non-metal manufacturing factories (25–35%), chemical factories (12–20%), and soil parent material (16–23%). The soil parent material sector stands out as the major contributor to Pb, reaching the largest contribution of 37–61%, while non-metal manufacturing factories had a significantly lower contribution, accounting for only 1–13%. Moreover, it was found that metal manufacturing factories were the primary source of As contamination, contributing 42–58%, while the contribution of soil parent material and non-metal manufacturing factories accounted for 14–22%, 10–17%.

$$\mathrm{Cd} = 0.65 \times (\mathrm{C_{AD}} \cdot \mathrm{S_{road}}^{(-\frac{2}{3})}) \cdot \mathrm{S_{NMMI}}^{(-\frac{2}{3})} + 528.81 \times (\mathrm{C_{AD}} \cdot \mathrm{S_{river}}^{(-\frac{5}{6})}) \cdot \mathrm{S_{CI}}^{(-\frac{5}{6})} + 1.58 \times (\mathrm{C_{AD}}^{(\frac{8}{5})} \cdot \mathrm{H}) \cdot \mathrm{S_{MI}}^{(-\frac{5}{8})} \quad R^2 = 0.670 \tag{5}$$

$$\mathrm{As} = 98.4 \times (\mathrm{C_{AD}}^{(\frac{1}{3})} \cdot \mathrm{S_{road}}^{(-\frac{5}{8})}) \cdot \mathrm{S_{NMMI}}^{(-\frac{5}{8})} + 0.054 \times (\mathrm{H}^{\frac{5}{2}} \cdot \mathrm{S_{river}}^{(-\frac{3}{4})}) \cdot \mathrm{S_{CI}}^{(-\frac{5}{6})} + 14815 \times (\mathrm{S_{road}}^{(-\frac{3}{4})} \cdot \mathrm{S_{river}}^{(-\frac{3}{4})}) \cdot \mathrm{S_{MI}}^{(-\frac{3}{4})} \quad R^2 = 0.651 \tag{6}$$

$$\mathrm{Pb} = 0.181 \times (\mathrm{C_{AD}} \cdot \mathrm{H}) \cdot \mathrm{S_{NMMI}}^{(-\frac{4}{5})} + 22247 \times (\mathrm{S_{road}}^{(-\frac{2}{3})} \cdot \mathrm{S_{river}}^{(-\frac{2}{3})}) \cdot \mathrm{S_{CI}}^{(-\frac{2}{3})} + 3.84 \times (\mathrm{H}^{\frac{3}{2}} \cdot \mathrm{S_{river}}^{(-\frac{3}{4})}) \cdot \mathrm{S_{MI}}^{(-\frac{3}{5})} \quad R^2 = 0.738 \tag{7}$$

where the subscript 'NMMI' represents the non-metal manufacturing factory; 'CI' means the chemical factory; 'MI' represented the metal manufacturing factory, and 'AD' represented the atmospheric deposition.

Table 2. Contributions from sources and their ranges, calculated by the three models.

	Element	NMMI	CI	MMI	Soil Parent Material
Model 1	Cd	25%	19%	33%	23%
	As	17%	24%	42%	17%
	Pb	13%	24%	17%	46%
Model 2	Cd	28%	12%	44%	16%
	As	11%	14%	61%	14%
	Pb	1%	17%	21%	61%
Model 3	Cd	35%	20%	30%	15%
	As	10%	11%	58%	22%
	Pb	7%	24%	32%	37%
Model 4	Cd	54%	1%	8%	37%
	As	33%	47%	17%	3%
	Pb	17%	32%	41%	10%
Range (Models 1, 2, 3)	Cd	25–35%	12–20%	30–44%	16–23%
	As	10–17%	11–24%	42–58%	14–22%
	Pb	1–13%	17–24%	17–32%	37–61%

Note: NMMI: non-metal manufacturing factory; CI: chemical factory; MMI: metal manufacturing factory.

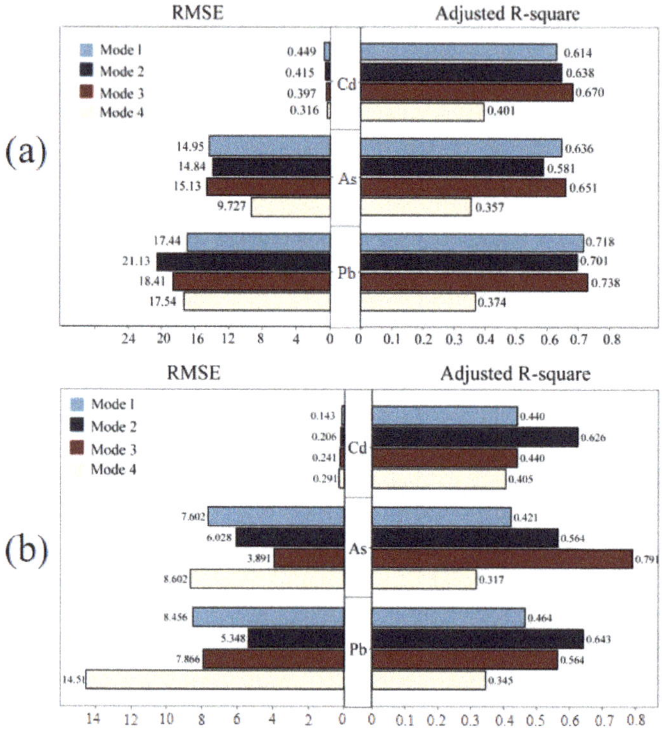

Figure 7. Comparison of fitting effect between four model fitting sets (**a**) and test sets (**b**).

3.5. Effect of Pollution Pathways on Heavy Metal Accumulation in Soil

The effect of the pollution pathway on soil heavy metals was quantified using the model obtained above (Figure 8). It was found that two factors, including altitude and atmospheric deposition, had a positive effect on heavy metals in soil. A change in the altitude level by 1 m resulted in a significant increase in the concentration of As and Cd, with values of 0.11 mg/kg and 0.003 mg/kg, respectively. It was also observed that an increment in concentration of Cd, Pb, and As in the atmospheric deposition by 0.1 μg/m^2/day led to elevated levels of the corresponding elements in the soil by 0.027 mg/kg, 0.016 mg/kg, and 0.04 mg/kg, respectively. In contrast, the three heavy metals displayed a nonlinear inverse relationship with both traffic emissions and irrigation water, whereas the magnitude of the effect diminished progressively as the distance increased. Specifically, for every 1000 m increase in road distance, the contents of Cd, Pb, and As in soil decreased by 0.034 mg/kg, 0.27 mg/kg, and 2.94 mg/kg, respectively. A similar trend was observed with the distance to rivers, showing decreasing heavy metals contents with an increase in distance to rivers, presenting values of 0.042 mg/kg/1000 m, 0.28 mg/kg/1000 m, and 2.11 mg/kg/1000 m, for Cd, Pb, and As, respectively. Meanwhile the amount of change in Pb increases instantly as the distance to the road increases to 1900 m. In general, the impact of roads and rivers progressively diminished and reached a state of stability as the distance extended beyond a specific threshold.

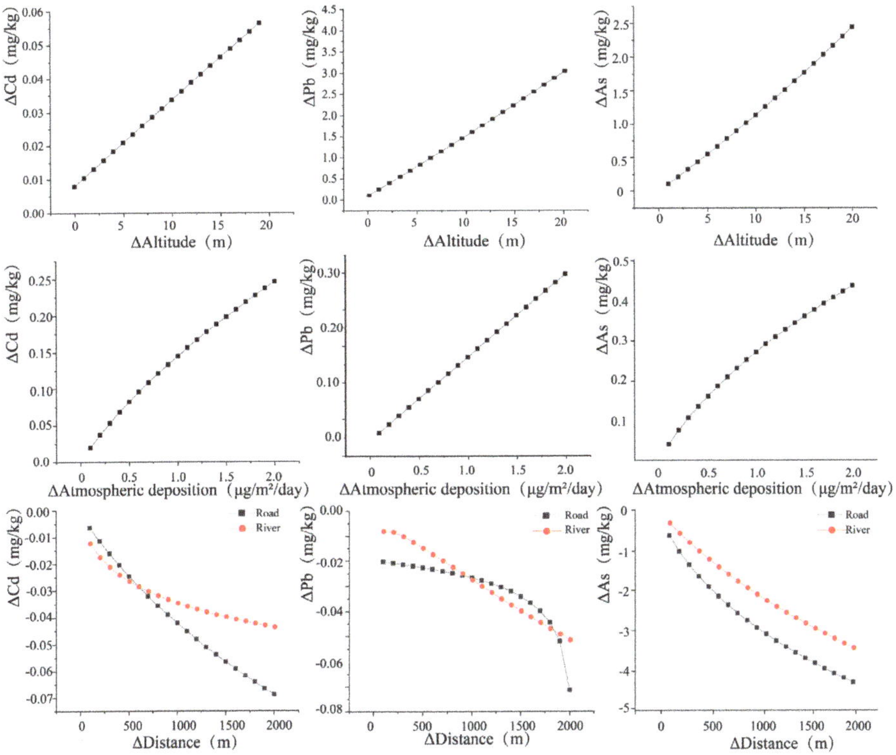

Figure 8. Effect of the amount of variation in different pollution pathway factors on soil heavy metals concentrations.

4. Discussion

The majority of the existing studies on pollution source analysis and identification primarily focused on the source type without identifying the actual situation of anthropogenic emissions. For example, Guan et al. (2018) employed a positive matrix factorization (PMF) model to identify the sources of soil heavy metals contamination in the Hexi Corridor region of China. In their assessment, the factors primarily explained by Cr and Ni were directly attributed to industrial sources [43]. However, the assessment technique demonstrates a significant level of subjectivity, and fails to adequately analyze the distinct categories of industrial sources. In addition, it is worth noting that other apportionment models such as principal component analysis and multiple linear regression (PCA-MLR) and Unmix also encounter similar problems [44,45]. Furthermore, a number of models can only characterize the sources into four broad categories, such as agriculture, factories, transportation, and natural sources, offering limited benefits for effectively controlling heavy metals in real practice. Therefore, our study aimed to improve the precision of specific sources, by considering the scale, type, and actual location of factories in the region. Instead of relying solely on the subjective assessment of key explanatory factors, the final analysis findings are better aligned with the actual pollution conditions in the study area.

During the source identification stage, the potential pollution sources were determined using GPCA along with the spatial distribution of the factories. The main explanatory factors of PCA1-1 included As, Cd, Pb, and Zn, and its factor scores interpolation depicted noteworthy similarities with the spatial distribution of chemical factories. Cd is a commonly used raw material in fireworks manufacturing. Pb has been identified as a distinctive component of traffic emissions in previous studies. However, with the prohibition of the usage of Pb in gasoline in China, traffic emission was no longer considered the main factor for Pb pollution [46]. The primary types of chemical factories in the study area include cement factories and fireworks factories. Previous studies have reported that As, Pb, and Cd in the raw materials of cement factories have the tendency to easily evaporate and release into the atmosphere alongside exhaust gases from cement kilns [13], while Zn is commonly used as a colorant in the production of firecrackers [31]. Further identification and comparison of pollution sources through GeoDetector found that the distance to the chemical plant had the highest degree of explanation for PCA1-1, and some studies have shown that the distance factor can reflect the degree of influence of pollution sources [47–49]. Therefore, the aforementioned conclusions can be combined to conclude that PCA1-1 serves as a source for chemical factories. The spatial distribution of PCA1-2 factor score interpolation showed similarities to the spatial distribution of metal manufacturing factories. The survey has revealed that the metal manufacturing factories in the study area are primarily composed of electroplating factories and steel factories. The main explanatory elements of PCA1-2 are As and Zn, which are widely used in the manufacturing of alloys, lead batteries, and semiconductor electronic devices to prepare stainless steel and corrosion-resistant alloys [44,50]. The results of GeoDetector also indicated that PCA1-2 corresponds to metal manufacturing factories. In the meantime, PCA2 mainly explained the presence of Cr and Ni, which are the main pollutants produced by garment and leather factories [32,33]. Similarly, the spatial distribution of non-metal manufacturing factories (garment factories) in the study area also showed similar characteristics to the factor scores of PCA2 in terms of the distribution of the high-value points. Therefore, PCA2 was identified as non-metal manufacturing factories. The main load elements of PCA3 are Ca and Mg and are considered to be unique elements of the soil matrix formation [51,52]. The simultaneous GeoDetector results indicate that soil type and land use type are highly explanatory for PCA3, thus identifying PCA3 as a source of soil-forming parent material.

Among the factor detector modules of the GeoDetector, it was observed that distance to roads had significant impact on As, which was further enhanced after the interaction with chemical factories. Therefore, it can be assumed that the accumulation of As in soil due to chemical factories is mainly associated with the transportation process. The results of the interaction detector module indicate that irrigation water served as the main contamination pathway for Cd and As emissions originating from non-metal manufacturing factories. The field survey revealed that the non-metal manufacturing factories were dominated by textile factories. The textile factories, being a significant contributor to wastewater, release substantial quantities of organic and inorganic pollutants into the river and atmosphere during the fabric manufacturing process [53–55]. Atmospheric deposition interaction analyses with non-metal manufacturing factories in the case of Cd, As and Pb showed a non-linear enhancement effect, which indicates that atmospheric deposition is the main pollution pathway from non-metal manufacturing factories in the study area. The above results on the pathways combined with the actual situation affirm the accuracy of the model in determining the potential sources of pollution. The PCA-MLRD model was employed to investigate the impact of pollution pathways on soil heavy metal concentrations. It was shown that when the distance from sampling sites to roads and rivers increased, the contents of heavy metals exhibited a significant exponential decline, which aligns with previous research [35,56]. Further comparisons between the PCA-MLRD model and the modified PCA-MLRD model revealed that the modified PCA-MLRD model had a greater accuracy, probably due to the following reasons: (1) Utilization of GPCA yielded a precise determination of types and locations of potential contamination sources, which accurately identified the actual sources of soil heavy metals contamination. (2) The original model only considered the location information of the contamination sources and relied only on distance as a representation of these contamination sources. However, the modified model added the factor of pollution pathways (traffic emissions, irrigation water, atmospheric deposition, and vertical migration) to the original model, which better simulated the contamination process of pollutants and ultimately enhanced the accuracy of results.

5. Conclusions

This study proposes a novel approach to heavy metal source apportionment in soil. Compared with the original PCA-MLRD, the modified model provided a more precise direction toward and basis for source identification. In addition to anthropogenic point sources, the model also accounted for non-point sources such as irrigation water, atmospheric emissions, and traffic emissions as pollution pathways. The accuracy of the model was improved, leading to more realistic results. The results indicated that metal factories had the highest contribution to Cd and As accumulations in soil, reaching 30–44% and 42–58%, respectively, and the highest contribution to Pb came from soil parent material, reaching 37–61%. The extent of the influence of pathway factors on soil heavy metals was also quantified. In terms of application scenarios, the model is more useful for promoting regional pollution control and precise management, specifically in areas with diverse pollution emission sources, such as urban–rural areas, or areas with diverse pollution pathways and dense river networks. Although the model needs further refining, this study provided a new perspective on source apportionment analysis for other researchers.

Supplementary Materials: The following supporting information can be downloaded at https://www.mdpi.com/article/10.3390/toxics12060382/s1, Table S1: Background values for different soil types.

Author Contributions: M.W.: Conceptualization, Methodology, Validation, Formal Analysis, Investigation, Writing—Original Draft. P.Y.: Investigation, Writing—Review and Editing. Z.T.: Methodology, Validation, Formal Analysis. X.S.: Methodology, Data Curation. J.P.: Supervision, Resources, Funding Acquisition. Y.H. (Yasir Hamid): Writing—Review and Editing. Y.H. (Ying Huang): Methodology, Validation, Formal Analysis, Writing—Review and Editing, Supervision, Funding Acquisition. All authors have read and agreed to the published version of the manuscript.

Funding: This work was financially supported by the Natural Science Foundation of Hunan Province, China (No. 2023JJ30305), China Postdoctoral Science Foundation (No. 2020M680782).

Institutional Review Board Statement: Not applicable.

Informed Consent Statement: Not applicable.

Data Availability Statement: Restrictions apply to the availability of these data.

Conflicts of Interest: The authors declare no conflicts of interest.

References

1. Shi, J.; Zhao, D.; Ren, F.; Huang, L. Spatiotemporal Variation of Soil Heavy Metals in China: The Pollution Status and Risk Assessment. *Sci. Total Environ.* **2023**, *871*, 161768. [CrossRef] [PubMed]
2. Wang, C.-C.; Zhang, Q.-C.; Yan, C.-A.; Tang, G.-Y.; Zhang, M.-Y.; Ma, L.Q.; Gu, R.-H.; Xiang, P. Heavy Metal(Loid)s in Agriculture Soils, Rice, and Wheat across China: Status Assessment and Spatiotemporal Analysis. *Sci. Total Environ.* **2023**, *882*, 163361. [CrossRef] [PubMed]
3. Hu, Y.; Cheng, H.; Tao, S. The Challenges and Solutions for Cadmium-Contaminated Rice in China: A Critical Review. *Environ. Int.* **2016**, *92–93*, 515–532. [CrossRef] [PubMed]
4. Liu, Y.; Cui, J.; Peng, Y.; Lu, Y.; Yao, D.; Yang, J.; He, Y. Atmospheric Deposition of Hazardous Elements and Its Accumulation in Both Soil and Grain of Winter Wheat in a Lead-Zinc Smelter Contaminated Area, Central China. *Sci. Total Environ.* **2020**, *707*, 135789. [CrossRef] [PubMed]
5. Huang, Y.; Li, T.; Wu, C.; He, Z.; Japenga, J.; Deng, M.; Yang, X. An Integrated Approach to Assess Heavy Metal Source Apportionment in Peri-Urban Agricultural Soils. *J. Hazard. Mater.* **2015**, *299*, 540–549. [CrossRef] [PubMed]
6. Huang, Y.; Deng, M.; Wu, S.; Japenga, J.; Li, T.; Yang, X.; He, Z. A Modified Receptor Model for Source Apportionment of Heavy Metal Pollution in Soil. *J. Hazard. Mater.* **2018**, *354*, 161–169. [CrossRef] [PubMed]
7. Zeng, W.; Wan, X.; Wang, L.; Lei, M.; Chen, T.; Gu, G. Apportionment and Location of Heavy Metal(Loid)s Pollution Sources for Soil and Dust Using the Combination of Principal Component Analysis, Geodetector, and Multiple Linear Regression of Distance. *J. Hazard. Mater.* **2022**, *438*, 129468. [CrossRef] [PubMed]
8. Zhao, R.; Guan, Q.; Luo, H.; Lin, J.; Yang, L.; Wang, F.; Pan, N.; Yang, Y. Fuzzy Synthetic Evaluation and Health Risk Assessment Quantification of Heavy Metals in Zhangye Agricultural Soil from the Perspective of Sources. *Sci. Total Environ.* **2019**, *697*, 134126. [CrossRef]
9. Chen, H.; Chai, M.; Cheng, J.; Wang, Y.; Tang, Z. Occurrence and Health Implications of Heavy Metals in Preschool Children's Clothing Manufactured in Four Asian Regions. *Ecotoxicol. Environ. Saf.* **2022**, *245*, 114121. [CrossRef]

10. Relić, D.; Sakan, S.; Anđelković, I.; Popović, A.; Đorđević, D. Pollution and Health Risk Assessments of Potentially Toxic Elements in Soil and Sediment Samples in a Petrochemical Industry and Surrounding Area. *Molecules* **2019**, *24*, 2139. [CrossRef]
11. Adnan, M.; Xiao, B.; Xiao, P.; Zhao, P.; Li, R.; Bibi, S. Research Progress on Heavy Metals Pollution in the Soil of Smelting Sites in China. *Toxics* **2022**, *10*, 231. [CrossRef] [PubMed]
12. Feng, W.; Guo, Z.; Xiao, X.; Peng, C.; Shi, L.; Ran, H.; Xu, W. Atmospheric Deposition as a Source of Cadmium and Lead to Soil-Rice System and Associated Risk Assessment. *Ecotoxicol. Environ. Saf.* **2019**, *180*, 160–167. [CrossRef] [PubMed]
13. Krishna, A.K.; Satyanarayanan, M.; Govil, P.K. Assessment of Heavy Metal Pollution in Water Using Multivariate Statistical Techniques in an Industrial Area: A Case Study from Patancheru, Medak District, Andhra Pradesh, India. *J. Hazard. Mater.* **2009**, *167*, 366–373. [CrossRef] [PubMed]
14. Yang, X.; Yang, Y.; Wan, Y.; Wu, R.; Feng, D.; Li, K. Source Identification and Comprehensive Apportionment of the Accumulation of Soil Heavy Metals by Integrating Pollution Landscapes, Pathways, and Receptors. *Sci. Total Environ.* **2021**, *786*, 147436. [CrossRef] [PubMed]
15. Zhang, Z.; Lu, Y.; Li, H.; Tu, Y.; Liu, B.; Yang, Z. Assessment of Heavy Metal Contamination, Distribution and Source Identification in the Sediments from the Zijiang River, China. *Sci. Total Environ.* **2018**, *645*, 235–243. [CrossRef] [PubMed]
16. Jafari, F.; Khademi, H. Spatial and Temporal Distribution of Heavy Metals Concentration in Atmospheric Dust in Kerman City. *J. Environ. Stud.* **2014**, *40*, 361–373. [CrossRef]
17. Sobhanardakani, S. Ecological and Human Health Risk Assessment of Heavy Metal Content of Atmospheric Dry Deposition, a Case Study: Kermanshah, Iran. *Biol Trace Elem Res* **2019**, *187*, 602–610. [CrossRef] [PubMed]
18. Zhu, Z.; Xu, Z.; Peng, J.; Fei, J.; Yu, P.; Wang, M.; Tan, Y.; Huang, Y.; Zhran, M.; Fahmy, A. The Contribution of Atmospheric Deposition of Cadmium and Lead to Their Accumulation in Rice Grains. *Plant Soil* **2022**, *477*, 373–387. [CrossRef]
19. Liao, S.; Jin, G.; Khan, M.A.; Zhu, Y.; Duan, L.; Luo, W.; Jia, J.; Zhong, B.; Ma, J.; Ye, Z.; et al. The Quantitative Source Apportionment of Heavy Metals in Peri-Urban Agricultural Soils with UNMIX and Input Fluxes Analysis. *Environ. Technol. Innov.* **2021**, *21*, 101232. [CrossRef]
20. Zhou, Y.; Jiang, D.; Ding, D.; Wu, Y.; Wei, J.; Kong, L.; Long, T.; Fan, T.; Deng, S. Ecological-Health Risks Assessment and Source Apportionment of Heavy Metals in Agricultural Soils around a Super-Sized Lead-Zinc Smelter with a Long Production History, in China. *Environ. Pollut.* **2022**, *307*, 119487. [CrossRef]
21. Zhou, J.; Du, B.; Liu, H.; Cui, H.; Zhang, W.; Fan, X.; Cui, J.; Zhou, J. The Bioavailability and Contribution of the Newly Deposited Heavy Metals (Copper and Lead) from Atmosphere to Rice (*Oryza sativa* L.). *J. Hazard. Mater.* **2020**, *384*, 121285. [CrossRef] [PubMed]
22. Azimi, S.; Cambier, P.; Lecuyer, I.; Thevenot, D. Heavy Metal Determination in Atmospheric Deposition and Other Fluxes in Northern France Agrosystems. *Water Air Soil Pollut.* **2004**, *157*, 295–313. [CrossRef]
23. Nicholson, F.A.; Smith, S.R.; Alloway, B.J.; Carlton-Smith, C.; Chambers, B.J. An Inventory of Heavy Metals Inputs to Agricultural Soils in England and Wales. *Sci. Total Environ.* **2003**, *311*, 205–219. [CrossRef] [PubMed]
24. Zhang, Y.; Zhang, S.; Zhu, F.; Wang, A.; Dai, H.; Cheng, S.; Wang, J.; Tang, L. Atmospheric Heavy Metal Deposition in Agro-Ecosystems in China. *Environ. Sci. Pollut. Res.* **2018**, *25*, 5822–5831. [CrossRef] [PubMed]
25. Yi, K.; Fan, W.; Chen, J.; Jiang, S.; Huang, S.; Peng, L.; Zeng, Q.; Luo, S. Annual Input and Output Fluxes of Heavy Metals to Paddy Fields in Four Types of Contaminated Areas in Hunan Province, China. *Sci. Total Environ.* **2018**, *634*, 67–76. [CrossRef]
26. Yu, E.; Liu, H.; Tu, Y.; Gu, X.; Ran, X.; Yu, Z.; Wu, P. Superposition Effects of Zinc Smelting Atmospheric Deposition on Soil Heavy Metal Pollution Under Geochemical Anomaly. *Front. Environ. Sci.* **2022**, *10*, 777894. [CrossRef]
27. Al-Khashman, O.A.; Shawabkeh, R.A. Metals Distribution in Soils around the Cement Factory in Southern Jordan. *Environ. Pollut.* **2006**, *140*, 387–394. [CrossRef]
28. Hua, S.; Tian, H.; Wang, K.; Zhu, C.; Gao, J.; Ma, Y.; Xue, Y.; Wang, Y.; Duan, S.; Zhou, J. Atmospheric Emission Inventory of Hazardous Air Pollutants from China's Cement Plants: Temporal Trends, Spatial Variation Characteristics and Scenario Projections. *Atmos. Environ.* **2016**, *128*, 1–9. [CrossRef]
29. Cao, X.; Zhang, X.; Tong, D.Q.; Chen, W.; Zhang, S.; Zhao, H.; Xiu, A. Review on Physicochemical Properties of Pollutants Released from Fireworks: Environmental and Health Effects and Prevention. *Environ. Rev.* **2018**, *26*, 133–155. [CrossRef]
30. Liu, Y.; Huang, H.; Sun, T.; Yuan, Y.; Pan, Y.; Xie, Y.; Fan, Z.; Wang, X. Comprehensive Risk Assessment and Source Apportionment of Heavy Metal Contamination in the Surface Sediment of the Yangtze River Anqing Section, China. *Environ. Earth Sci.* **2018**, *77*, 493. [CrossRef]
31. Chen, Y.-F.; Shi, Q.-Y.; Qu, J.-Y.; He, M.-X.; Liu, Q. A Pollution Risk Assessment and Source Analysis of Heavy Metals in Sediments: A Case Study of Lake Gehu, China. *Chin. J. Anal. Chem.* **2022**, *50*, 100077. [CrossRef]
32. Sanyal, T.; Kaviraj, A.; Saha, S. Deposition of Chromium in Aquatic Ecosystem from Effluents of Handloom Textile Industries in Ranaghat–Fulia Region of West Bengal, India. *J. Adv. Res.* **2015**, *6*, 995–1002. [CrossRef]
33. Ahsan, M.A.; Satter, F.; Siddique, M.A.B.; Akbor, M.A.; Ahmed, S.; Shajahan, M.; Khan, R. Chemical and Physicochemical Characterization of Effluents from the Tanning and Textile Industries in Bangladesh with Multivariate Statistical Approach. *Environ. Monit. Assess.* **2019**, *191*, 575. [CrossRef] [PubMed]
34. Qiang, L.; Yang, W.; Jingshuang, L.; Quanying, W.; Mingying, Z. Grain-Size Distribution and Heavy Metal Contamination of Road Dusts in Urban Parks and Squares in Changchun, China. *Environ. Geochem. Health* **2015**, *37*, 71–82. [CrossRef] [PubMed]

35. Yan, G.; Mao, L.; Jiang, B.; Chen, X.; Gao, Y.; Chen, C.; Li, F.; Chen, L. The Source Apportionment, Pollution Characteristic and Mobility of Sb in Roadside Soils Affected by Traffic and Industrial Activities. *J. Hazard. Mater.* **2020**, *384*, 121352. [CrossRef]
36. Yuan, Z.; Luo, T.; Liu, X.; Hua, H.; Zhuang, Y.; Zhang, X.; Zhang, L.; Zhang, Y.; Xu, W.; Ren, J. Tracing Anthropogenic Cadmium Emissions: From Sources to Pollution. *Sci. Total Environ.* **2019**, *676*, 87–96. [CrossRef] [PubMed]
37. Liu, P.; Wu, Q.; Hu, W.; Tian, K.; Huang, B.; Zhao, Y. Effects of Atmospheric Deposition on Heavy Metals Accumulation in Agricultural Soils: Evidence from Field Monitoring and Pb Isotope Analysis. *Environ. Pollut.* **2023**, *330*, 121740. [CrossRef] [PubMed]
38. Wang, J.; Li, X.; Christakos, G.; Liao, Y.; Zhang, T.; Gu, X.; Zheng, X. Geographical Detectors-Based Health Risk Assessment and Its Application in the Neural Tube Defects Study of the Heshun Region, China. *Int. J. Geogr. Inf. Sci.* **2010**, *24*, 107–127. [CrossRef]
39. Su, Y.; Li, T.; Cheng, S.; Wang, X. Spatial Distribution Exploration and Driving Factor Identification for Soil Salinisation Based on Geodetector Models in Coastal Area. *Ecol. Eng.* **2020**, *156*, 105961. [CrossRef]
40. Huang, S.; Xiao, L.; Zhang, Y.; Wang, L.; Tang, L. Interactive Effects of Natural and Anthropogenic Factors on Heterogenetic Accumulations of Heavy Metals in Surface Soils through Geodetector Analysis. *Sci. Total Environ.* **2021**, *789*, 147937. [CrossRef]
41. Tao, H.; Liao, X.; Li, Y.; Xu, C.; Zhu, G.; Cassidy, D.P. Quantifying Influences of Interacting Anthropogenic-Natural Factors on Trace Element Accumulation and Pollution Risk in Karst Soil. *Sci. Total Environ.* **2020**, *721*, 137770. [CrossRef]
42. Kaiser, H.F. The Application of Electronic Computers to Factor Analysis. *Educ. Psychol. Meas.* **1960**, *20*, 141–151. [CrossRef]
43. Guan, Q.; Wang, F.; Xu, C.; Pan, N.; Lin, J.; Zhao, R.; Yang, Y.; Luo, H. Source Apportionment of Heavy Metals in Agricultural Soil Based on PMF: A Case Study in Hexi Corridor, Northwest China. *Chemosphere* **2018**, *193*, 189–197. [CrossRef]
44. Chen, Z.; Ding, Y.; Jiang, X.; Duan, H.; Ruan, X.; Li, Z.; Li, Y. Combination of UNMIX, PMF Model and Pb-Zn-Cu Isotopic Compositions for Quantitative Source Apportionment of Heavy Metals in Suburban Agricultural Soils. *Ecotoxicol. Environ. Saf.* **2022**, *234*, 113369. [CrossRef] [PubMed]
45. Su, C.; Meng, J.; Zhou, Y.; Bi, R.; Chen, Z.; Diao, J.; Huang, Z.; Kan, Z.; Wang, T. Heavy Metals in Soils from Intense Industrial Areas in South China: Spatial Distribution, Source Apportionment, and Risk Assessment. *Front. Environ. Sci.* **2022**, *10*, 820536. [CrossRef]
46. Sun, Y.; Zhuang, G.; Zhang, W.; Wang, Y.; Zhuang, Y. Characteristics and Sources of Lead Pollution after Phasing out Leaded Gasoline in Beijing. *Atmos. Environ.* **2006**, *40*, 2973–2985. [CrossRef]
47. Chanchpara, A.; Muduli, M.; Prabhakar, V.; Madhava, A.K.; Thorat, R.B.; Haldar, S.; Ray, S. Pre-to-Post Diwali Air Quality Assessment and Particulate Matter Characterization of a Western Coastal Place in India. *Environ. Monit. Assess.* **2023**, *195*, 413. [CrossRef] [PubMed]
48. Zhang, P.; Qin, C.; Hong, X.; Kang, G.; Qin, M.; Yang, D.; Pang, B.; Li, Y.; He, J.; Dick, R.P. Risk Assessment and Source Analysis of Soil Heavy Metal Pollution from Lower Reaches of Yellow River Irrigation in China. *Sci. Total Environ.* **2018**, *633*, 1136–1147. [CrossRef]
49. Qiao, P.; Lei, M.; Yang, S.; Yang, J.; Guo, G.; Zhou, X. Comparing Ordinary Kriging and Inverse Distance Weighting for Soil as Pollution in Beijing. *Environ. Sci. Pollut. Res.* **2018**, *25*, 15597–15608. [CrossRef]
50. Huaiwei, Z.; Xin, H. An Overview for the Utilization of Wastes from Stainless Steel Industries. *Resour. Conserv. Recycl.* **2011**, *55*, 745–754. [CrossRef]
51. Bhuiyan, M.A.H.; Parvez, L.; Islam, M.A.; Dampare, S.B.; Suzuki, S. Heavy Metal Pollution of Coal Mine-Affected Agricultural Soils in the Northern Part of Bangladesh. *J. Hazard. Mater.* **2010**, *173*, 384–392. [CrossRef] [PubMed]
52. Luo, Y.; Shi, C.; Yang, S.; Liu, Y.; Zhao, S.; Zhang, C. Characteristics of Soil Calcium Content Distribution in Karst Dry-Hot Valley and Its Influencing Factors. *Water* **2023**, *15*, 1119. [CrossRef]
53. Dung, T.T.T.; Cappuyns, V.; Swennen, R.; Phung, N.K. From Geochemical Background Determination to Pollution Assessment of Heavy Metals in Sediments and Soils. *Rev. Environ. Sci. Biotechnol.* **2013**, *12*, 335–353. [CrossRef]
54. Bilińska, L.; Gmurek, M.; Ledakowicz, S. Comparison between Industrial and Simulated Textile Wastewater Treatment by AOPs—Biodegradability, Toxicity and Cost Assessment. *Chem. Eng. J.* **2016**, *306*, 550–559. [CrossRef]
55. Berthiaume, A.; Monette, A.; Rosenberger, J.; Lupien, B. Comment on Giacosa et al. Characterization of Annual Air Emissions Reported by Pulp and Paper Mills in Atlantic Canada. Pollutants **2022**, 2, 135–155. *Pollutants* **2022**, *2*, 328–329. [CrossRef]
56. Liu, Z.; Fei, Y.; Shi, H.; Mo, L.; Qi, J. Prediction of High-Risk Areas of Soil Heavy Metal Pollution with Multiple Factors on a Large Scale in Industrial Agglomeration Areas. *Sci. Total Environ.* **2022**, *808*, 151874. [CrossRef]

toxics

MDPI

Article

Temporal and Spatial Variation of Toxic Metal Concentrations in Cultivated Soil in Jiaxing, Zhejiang Province, China: Characteristics and Mechanisms

Mengzhuo Cao [1,4], Yanbo Jia [2], Xin Lu [1], Jinfa Huang [3], Yanlai Yao [1], Leidong Hong [1], Weijing Zhu [1], Weiping Wang [1], Fengxiang Zhu [1] and Chunlai Hong [1,*]

1 Institute of Environment, Resource, Soil and Fertilizer, Zhejiang Academy of Agricultural Sciences, Hangzhou 310021, China; cmz201309@163.com (M.C.); q2080252919@163.com (X.L.); yaoyl@zaas.an.cn (Y.Y.); hongld@zaas.ac.cn (L.H.); zhuwj@zaas.ac.cn (W.Z.); wangwp@zaas.ac.cn (W.W.); zhufx@zaas.ac.cn (F.Z.)
2 Hangzhou Institution of Food and Drug Control, Hangzhou 310022, China; jiaboshi2002@sina.com
3 Jiaxing Soil and Fertilizer Plant Protection and Rural Energy Station, Jiaxing Agriculture and Rural Bureau, Jiaxing 314050, China; jxstfz@126.com
4 Shanghai Huadi Environmental Technology Co., Ltd., Shanghai 201803, China
* Correspondence: hongcl@zaas.ac.cn; Tel.: +0571-86409538

Abstract: The toxic metal (As, Cd, Cr, Cu, Hg, Ni, Pb, and Zn) pollution in 250 agricultural soil samples representing the urban area of Jiaxing was studied to investigate the temporal and spatial variations. Compared to the early 1990s, the pollution level has increased. Industry and urbanization were the main factors causing toxic metal pollution on temporal variation, especially the use of feed containing toxic metals. The soil types and crop cultivation methods are the main factors causing toxic metal pollution on spatial variation. Although the single-factor pollution indices of all the toxic metals were within the safe limits, as per the National Soil Environmental Quality Standard (risk screening value), if the background values of soil elements in Jiaxing City are used as the standard, the pollution index of all the elements surveyed exceeds 1.0, reaching a level of mild pollution. The soil samples investigated were heavily contaminated with toxic metal compounds, and their levels increased over time. This situation poses potential ecological and health risks.

Keywords: agriculture; concentration; safe limits; national risk screening value; pollution

1. Introduction

In recent years, intensive urbanization, industrialization, and agricultural activities have had negative effects on farmland and agricultural production. In particular, contamination of soil with toxic metals in urban regions is posing a threat to the safety of agricultural products [1–3]. Toxic metal pollution has the characteristics of universality, concealment, surface aggregation, and irreversibility. Therefore, the pollution caused by toxic metals in farmland soil is not easy to find and solve. Based on this fact, the investigation and evaluation of toxic metal accumulation in farmland soil have been widely concerned [3].

China is an agriculturally large country with an arable area comprising 8% of the world's total arable area and a high food production of 500 million tons per annum [4]. Jiaxing is located in the Hang-Jia-Hu plain, which is the most important agricultural production base in the Zhejiang Province of China. The farmland area in Jiaxing is less than 14% of Zhejiang Province, but it produces 30% and 48% of the total yield of vegetables and grains, respectively, in Zhejiang. Therefore, the soil quality in Jiaxing is of great significance for the safety of agricultural products and human health in Zhejiang.

In recent years, some researchers have reported the toxic metal pollution of soil and agricultural products in Jiaxing [4,5]. Liu et al. assessed the heavy metal pollution in the soil around a plastic metal factory in Jiaxing and found that the ecological risk was great [6].

Hu et al. analyzed the spatiotemporal variation and temporal changes in the sources of Cr, Pb, Cd, Hg, and As in the soil of Jiaxing based on 4359 soil samples collected in 2002 and 2012 [7]. However, the studies were local and sporadic, so the toxic metal pollution status of the soil in the entire region and the temporal and spatial variation remained unclear. The mechanism of temporal and spatial changes in toxic metal concentrations in soil still needs to be explored. The main objective of this research was to evaluate the toxic metal pollution of soil in Jiaxing and compare the local accumulation and temporal variation characteristics with the survey monitoring data of the early 1990s and the National Soil Environmental Quality Standard (GB15618-2018). The findings of the study could provide comprehensive information on the soil quality in this region and help to design strategies to minimize toxic metal transfer in the food chain.

2. Materials and Methods

2.1. Study Area

Jiaxing is located in the center of Shanghai, Hangzhou, and Suzhou, as well as other middle- to large-sized cities. It has jurisdiction over five counties (cities), including Haining, Tongxiang, Haiyan, Pinghu, and Jiashan, as well as the two districts of Nanhu and Xiuzhou, and has a total land area of 3915 km^2.

Regarding the climate, Jiaxing belongs to the East Asian monsoon region and has four distinct seasons. Its annual average temperature is 15.9 °C, precipitation is 1168.6 mm/year, and sunshine is 2017.0 h/year. Because of the low-lying terrain, the soil in Jiaxing consists of shallow marine sediments, river alluvium, lacustrine, and paludal sediments [8]. The soil types mainly include paddy, fluvo-aquic, and coastal saline soils. Paddy soil includes silt-clayey yellow mottled, blue clayey, and silt-loamy types [7].

2.2. Soil Sampling and Processing

Sites were selected at one point per 1/15 hectare with a 1:50,000 soil map, and the Global Positioning System was used to position each sampling point. A representative plot was selected at each point, and soil samples were collected using a soil sampler in the tillage layer at a depth of 0–20 cm. Each representative sample from a point consisted of a quincunx multipoint sample (taken from at least ten sub-points) [9]. About 1 kg of each evenly mixed representative soil sample was taken for testing. In total, 250 soil samples were collected from the entire study area in the suburbs of Jiaxing (Figure 1).

Soil samples were air-dried and screened through a 100-mesh sieve. To determine the concentration of toxic metals, the HNO_3-$HClO_4$-HF digestion method was used to prepare the samples. Available forms of various toxic metals (As, Cd, Cr, Cu, Hg, Ni, Pb, and Zn) were extracted with 0.1 mol ethylene diaminetetraacetic acid (EDTA) solution (pH 7.0) from soil–water suspensions prepared with a ratio of 1:5. The contents of As and Hg in the samples were determined through cold atomic absorption spectrophotometry (3400 ASS, Thermo Scientific, Waltham, Massachusetts, USA) [10]. The contents of Pb, Cu, Zn, Cr, Ni, and Cd were measured by the guidelines for graphite furnace atomic absorption spectrometry (240Z AA, Agilent, Palo Alto, California, USA) [11]. During all the analytical determinations, the detection limits for As, Cd, Cr, Cu, Hg, Ni, Pb, and Zn were 0.002 mg/kg, 0.01 mg/kg, 1.2 mg/kg, 0.01 mg/kg, 0.002 mg/kg, 0.3 mg/kg, 0.1 mg/kg, and 0.008 mg/kg, respectively. In addition, some national standard soil samples were also analyzed to ensure quality control.

Soil pH was measured by potentiometry [12]. The contents of organic matter in soil samples were determined by titrimetry using sulfuric acid, potassium dichromate, and ferrous sulfate after thermal oxidation [13].

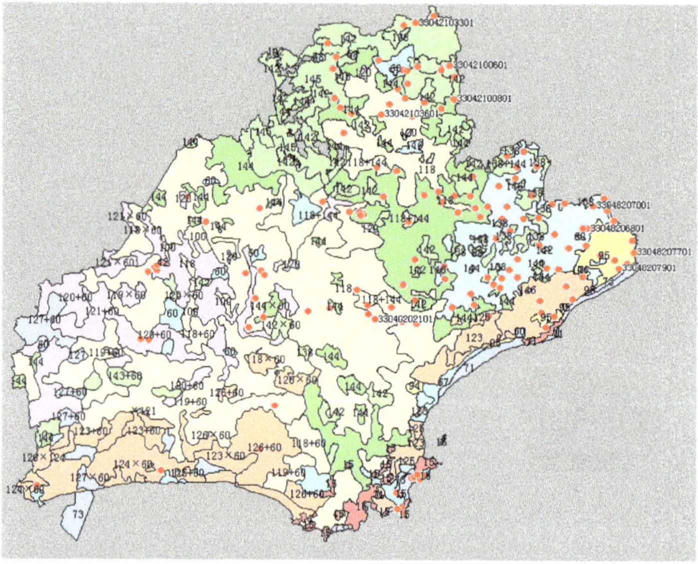

Figure 1. Distribution map of heavy metal elements in Jiaxing soil. The planar diagram of soil sampling locations in Jiaxing City. Red circles denote sampling points, and the numbers represent the respective point identifiers.

2.3. Data Processing and Statistical Analysis

The average values, coefficients of variation, normal distribution test, and other described statistical analyses of the toxic metal contents were performed by the SPSS 12.0 (SPSS, Chicago, IL, USA) statistical software. Variogram fitting and relevant parameters were obtained using the Statistical Analysis System (SAS 9.3, Cary, North Carolina, USA).

2.4. Evaluation Method

The criterion used to evaluate soil quality was the risk screening limits of the National Environmental Quality Standard for Soils. While evaluating the concentration of a single toxic metal in soil, the fuzzy synthetic evaluation method was used to determine the overall pollution level of the soil with toxic metals [14].

2.4.1. Single-Factor Pollution Evaluation

The formula used to calculate the single-factor pollution index is as follows [15]:

$$Pi = Ci/Si,$$

where Pi is the pollution index of pollutant i, Ci is the measured concentration of the pollution index of pollutant i, and Si is the evaluation criterion of pollutant i. The Pi values for different pollution categories are as follows: Pi $\leq$0.7 = excellent, Pi $\leq$1.0 = safe, Pi $\leq$2 = lightly polluted, Pi $\leq$3 = medium-level pollution/moderately polluted, and Pi >3 = heavily polluted.

2.4.2. Comprehensive Pollution Index

The single-factor index only reflects the pollution level of a single pollutant rather than the integrated pollution level of soil and crops. In contrast, the comprehensive pollution index takes into account the average and highest values of the single-factor pollution index and can highlight the role of pollutants causing serious pollution. The comprehensive pollution index is calculated as follows [16]:

$$P_{com} = \{[(Ci/Si)^2{}_{max} + (Ci/Si)^2{}_{ave}]/2\}^{1/2},$$

where P_{com} is the comprehensive pollution index, Pi = Ci/Si, P_{ave} is the average value of the various pollution index Pi, and P_{max} is the maximum value among the individual pollution indices. The P_{com} values for pollution categories are as follows: $P_{com} \leq 0.7$ = safe, $P_{com} \leq 1.0$ = warning line, $P_{com} \leq 2.0$ = lightly polluted, $P_{com} \leq 3.0$ = medium-level pollution/moderately polluted, and $P_{com} > 3.0$ = heavy pollution/heavily polluted.

3. Results and Discussions

3.1. Temporal and Spatial Variation of Toxic Metal Pollution in Soil

3.1.1. Temporal Variation of Toxic Metal Pollution in Soil

According to Figure 2A, with the risk screening value of the National Soil Environmental Quality Standard (GB15618-2018) as the indicator, the unit pollution indices of all the toxic metals are less than 1.0. The single-factor pollution indices of all of the elements are in the safe range, in which the "excellent" category (with a pollution index less than 0.70) is up to 50%. More than 0.7 and less than 1.0 are the warning ranges. The single pollution indices of Cd, Hg, Ni, Cu, and As exceeded the warning value of 0.7 and were less than 1.0. According to Figure 2B, the proportions of Ni, Hg, Cd, Cu, and As with single pollution indices of more than 0.7 and less than 1.0 in all soil samples were 25.6%, 18.8%, 11.2%, 2%, and 0.4%, respectively. However, no sample was over-detected for Pb, Zn, or Cr.

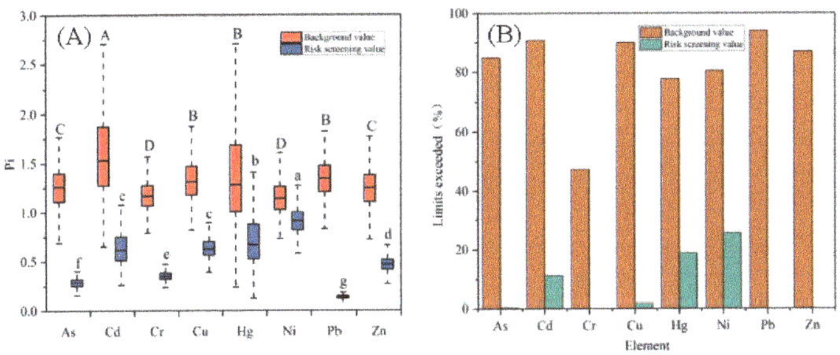

Figure 2. Evaluation of toxic metal element pollution in the soils of Jiaxing City. (**A**) Soil toxic metal single-factor pollution index in Jiaxing. (**B**) Excess rate of toxic metals in soil in Jiaxing City. Uppercase letters denote significance based on background values, while lowercase letters indicate significance based on risk values ($p < 0.05$).

However, according to Figure 2A, if the background values of soil elements in Jiaxing City were used as a standard, the pollution indices of all the elements investigated exceeded 1.0 and reached a mild pollution level. Among them, Cd and Hg contaminated the most seriously, reaching more than 1.5-fold of the background value. The excess rates of the metallic elements were greater than 44.6% using the local elemental background as the evaluation standard, among which Pb, CD, and Cu were over 90%. Comparing recent values of soil toxic metal tests with those obtained in Jiaxing City in 1981 (Table 1), only As and Hg levels in the soil did not increase. Cd, Zn, Cu, Pb, Ni, and Cr increased by 86.2%, 27.2%, 41.1%, 45.6%, 4.94%, and 21.1%, respectively.

Table 1. Comparison of Soil Environmental Quality Evaluation Standards.

Item	As	Cd	Cr	Cu	Hg	Ni	Pb	Zn
Risk screening value (μg/g)	30	0.300	250	50	0.250	40	250	200
Background value (μg/g) *	6.8	0.120	74.9	23.8	0.130	31.8	25.1	74.8
Soil test values in 1981 (μg/g) **	9.87	0.116	72.34	22.67	0.195	34.59	23.08	74.11

μg/g is the quality of toxic metals in the tested soil. "*" is the background value of soil toxic metal concentration in Hangjiahu plain in the early 1990s; "**" is the background content of toxic metals obtained by the Jiaxing environmental monitoring station after sampling 28 points in Jiaxing, Jiashan, Pinghu, Haiyan, Haining, and Tongxiang counties (cities) in 1981.

Toxic metal composite contamination was severe in soils from the investigated sample sites. A total of 92.6% of the sample points had composite pollution, and the percentage of sample points with the eight toxic metals represented all over reached 16.7%. The above results demonstrated that a significant accumulation of toxic metals was occurring in the form of composite pollutants in the soils of Jiaxing City. This is also indicated by the correlation analysis in Figure 3. Cd, Cr, Cu, Ni, Pb, and Zn showed highly significant correlations with each other. This further illustrates that the above six elements are composite pollutants and that their sources share homology. The toxic metal contamination of soils is mainly composite pollution. With the single-factor pollution index (Pi) alone, it is difficult to make an objective evaluation of the overall pollution situation of soils. However, the comprehensive pollution index is able to make up for the shortage of the single-factor pollution index. With two different evaluation criteria, the calculated comprehensive pollution index was quite different. Calculations using the national secondary standard for soil environmental quality yielded a comprehensive pollution index of 0.837, indicating that the soil environmental quality in Jiaxing City is still in a safety vigilance state. However, using the background values of soil elements in Jiaxing City as the standard, the obtained comprehensive pollution index reached 1.859, indicating that the soil environment quality in Jiaxing City had reached a slightly polluted level.

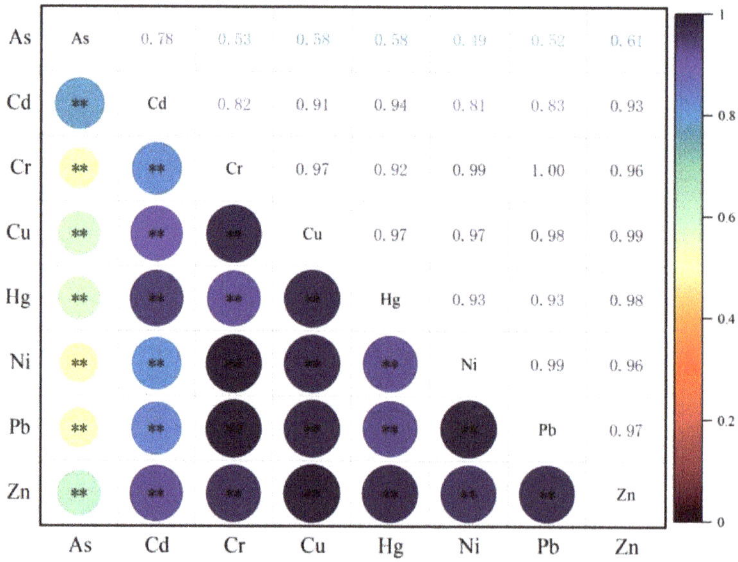

Figure 3. Correlation analysis of toxic metal elements in the soils of Jiaxing City. Pearson's statistical correlation analysis was employed, where deeper colors indicate stronger correlations. (** $p < 0.01$).

3.1.2. Spatial Variation of Toxic Metal Pollution

As shown in Table 2, in the soil samples from Jiaxing, the concentrations of As, Cd, and Hg varied considerably in different soil types, and the maximum values were more than 10 times higher than the minimum values. However, the contents of several other elements in different soil samples were relatively close, and the maximum value was 2–5 times the minimum value. The coefficient of variation reflects the average degree of variation between different sample points. Table 2 shows that the coefficients of variation of As, Cd, and Hg were the largest at over 0.50. The results showed large variability and spatial differentiation, indicating that outside interference considerably changed the contents of these elements in the soil. These differences might be attributed to anthropogenic activities, farming, management practices, and cropping systems [17,18].

Table 2. Descriptive statistical analysis of soil total toxic metals in Jiaxing City (μg/g).

Elements	Mean	STD	Max.	Min.	Variation Coefficient	Skewness	Kurtosis
As	8.76	4.63	77.33	4.07	0.53	13.11	194.63
Cd	0.22	0.16	1.26	0.078	0.53	4.29	30.08
Cr	87.56	11.75	125.59	55.73	0.13	−0.061	0.19
Cu	32.03	6.82	61.72	16.58	0.21	0.94	2.49
Hg	0.19	0.098	0.65	0.031	0.51	1.84	4.87
Ni	36.31	5.77	51.01	23.20	0.16	0.11	−0.18
Pb	33.62	5.48	50.76	16.61	0.16	0.053	0.51
Zn	94.28	18.50	188.21	53.80	0.19	1.22	4.27

Some elements had smaller coefficients of variation (all around 0.20), suggesting that these elements were subject to consistent external influences. The spatial differentiation is low, suggesting that these elements may have the same origin in this region. As can be seen from Table 3 and Figure 4, there were some differences in the levels of toxic metals in different counties, municipalities, and districts of Jiaxing City. The main areas with higher pollution were the South Lake area and Jiashan County. The single-factor pollution indices reached 1.05–1.67 and 1.0–1.77, respectively, using the background values of toxic metals in soils from Jiaxing City as the standard. The most serious pollution of Cd, Cu, Hg, Pb, and Zn was in Jiashan County, with single-factor pollution indices of 1.06–2.82, 1.26–1.65, 0.93–2.50, 1.21–1.69, and 1.03–1.77, respectively. The pollution index of each toxic metal was relatively low in the Xiuzhou region. This trend was also reflected in the cases of exceedance rates at different localities (Figure 5). Using the background values from soils in Jiaxing City as a standard, the Cr exceedance rate in Jiashan County was 80%, while all the remaining elements exceeded 90%. The exceedances of As, Cd, Cr, Cu, Pb, and Zn in the South Lake region reached 100%, and the overall exceedances in other regions were all above 50%. The accumulation of each toxic metal element in different counties, municipalities, and regions has led to a large difference in the comprehensive pollution index in soils. The three regions with the largest comprehensive pollution index were Jiashan County, South Lake area, and Tongxiang City. Its comprehensive pollution index was calculated using the background value as the standard, and the obtained values were 2.18, 2.02, and 1.97. The obtained values were 0.92, 0.90, and 0.89, calculated using the secondary standard for soil environmental quality. The comprehensive pollution index of Xiuzhou District was the smallest, which may be related to the collection of a certain amount of pear garden soil, which was relatively small due to exogenous pollution, while the soil environmental quality remained relatively good.

Table 3. Statistical analysis of soil toxic metal elements in counties (μg/g).

Element		Nanhu	Xiuzhou	Jiashan	Haiyan	Haining	Pinghu	Tongxiang
As	Content range	7.2–12.5	4.6–10.7	6.8–12.1	6.1–10.6	5.7–12.0	4.1–11.2	5.1–77.3
	Mean	9.2	8.1	9.7	8.1	7.9	8.2	11.7
Cd	Content range	0.1–0.7	0.1–0.4	0.1–1.3	0.1–0.5	0.1–0.5	0.1–0.8	0.1–0.4
	Mean	0.2	0.2	0.3	0.2	0.2	0.2	0.2
Cr	Content range	76.8–109.3	66.8–112.7	57.1–125.6	59.1–117.6	55.7–109.6	62.5–106.5	70.6–106.3
	Mean	89.2	87.4	84.3	92.7	83.0	87.0	89.3
Cu	Content range	25.0–43.8	20.5–42.1	22.4–61.7	21.3–39.9	16.5–35.5	16.8–53.3	20–42.3
	Mean	34.4	29.1	37.9	31.1	26.9	30.9	29.3
Hg	Content range	0.1–0.5	0.1–0.5	0.1–0.7	0.1–0.2	0.1–0.3	0.1–0.5	0.1–0.5
	Mean	0.2	0.1	0.3	0.2	0.2	0.2	0.2
Ni	Content range	31.6–48.9	26.0–48.5	26.0–46.4	25.0–51.0	23.4–48.7	23.2–48.3	29.7–49.9
	Mean	37.7	36.1	36.5	38.6	34.0	34.7	38.0
Pb	Content range	26.2–48.3	22.4–49.4	22.0–42.3	22.9–39.9	16.6–39.7	17.6–41.1	23.2–44.2
	Mean	34.5	31.7	36.2	33.7	30.7	32.6	34.0
Zn	Content range	77.1–132.4	59.2–117.4	62.1–134.9	58.2–118.9	60.2–106.2	53.8–188.2	64.3–162.1
	Mean	99.1	87.0	103.5	90.6	84.4	92.0	93.5

The number of soil samples from Nanhu, Xiuzhou, Jiashan, Haiyan, Haining, Pinghu, and Tongxiang were 33, 28, 19, 20, 30, 80, and 40, respectively.

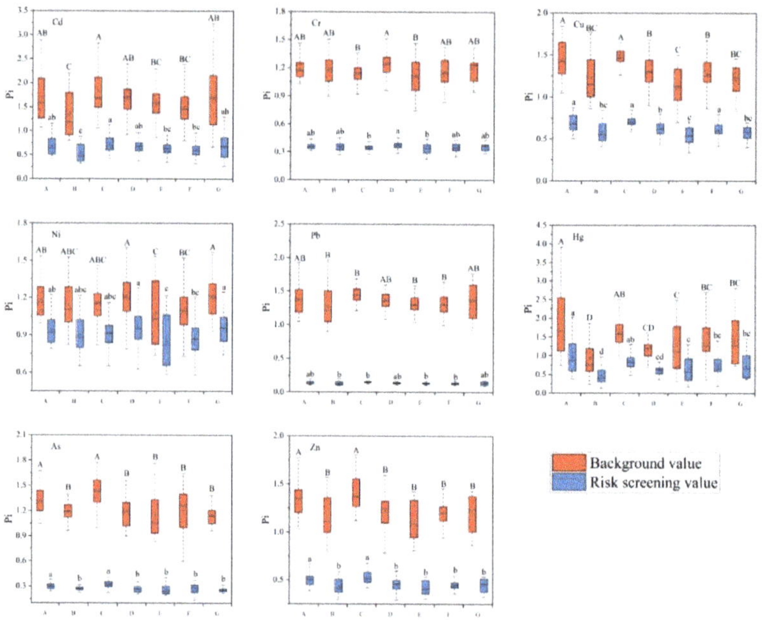

Figure 4. Evaluation of toxic metal element pollution in soils of different regions. (A) Nanhu District. (B) Xiuzhou District. (C) Jiashan District. (D) Haiyan District. (E) Haining District. (F) Pinghu District. (G) Tongxiang District. Uppercase letters denote significance based on background values, while lowercase letters indicate significance based on risk values ($p < 0.05$).

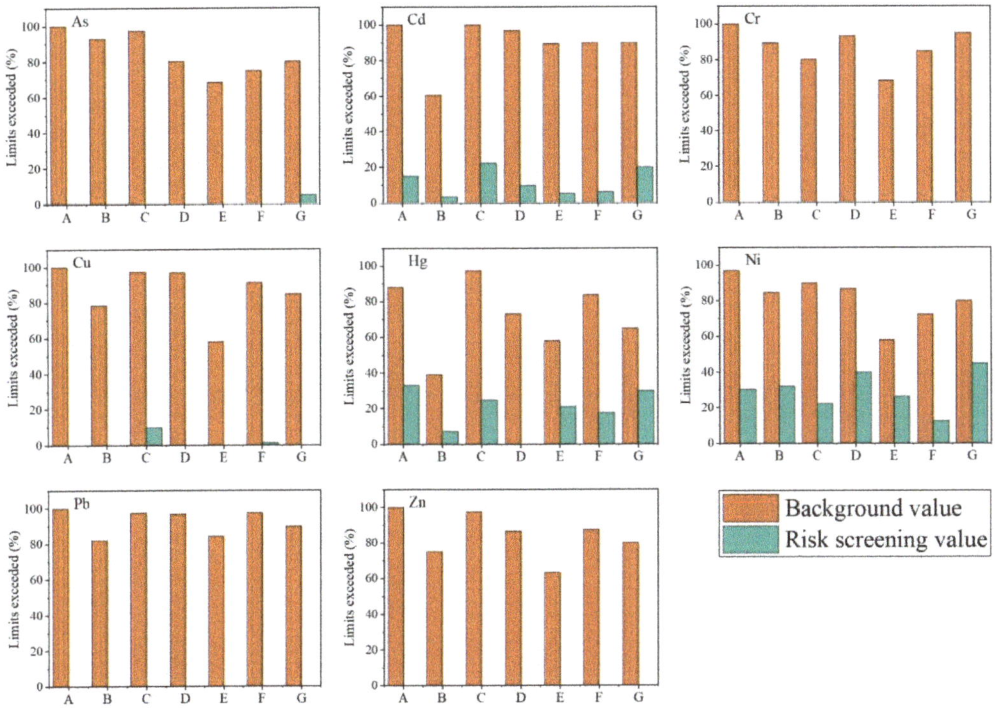

Figure 5. Excess rate of toxic metals in the soil in different regions. (A) Nanhu District. (B) Xiuzhou District. (C) Jiashan District. (D) Haiyan District. (E) Haining District. (F) Pinghu District. (G) Tongxiang District.

3.2. Mechanism of Temporal and Spatial Changes in Toxic Metal Concentrations in the Soil

3.2.1. Temporal Variation of Toxic Metals Affected by People's Production

The degree of toxic metal pollution in agricultural soil is closely related to the level of regional economic development, especially the level of industry, transportation, and urbanization [19,20]. According to previous studies, first of all, industrial sources are the main reason for the accumulation of Cu, Zn, and Pb in the soil, such as the mining of minerals and the random disposal of wastes in industrial production [21]. Secondly, inappropriate agricultural measures, including sewage irrigation and fertilizer application, also become the primary causes of soil toxic metal pollution in cities with developed agriculture. The lack of water resources makes sewage an important part of irrigation water. Sewage irrigation brings a majority of harmful substances (e.g., Cd, Cr, and Zn), causing different levels of soil pollution [22]. Thirdly, vehicle exhaust emissions will significantly increase the concentration of Pb and Zn on both sides of the road [23]. Fourthly, with the development of urbanization, more household garbage will be produced. The metabolism and degradation of harmful substances are relatively slow in a closed ecosystem. Compared with the natural environment, toxic metal pollution in the urban environment is more likely to become serious [24]. In this study, with the industrialization and urbanization of Jiaxing arable land, soil toxic metal pollution has increased in the past two decades. The main pollutants are Cd, Cu, Zn, etc., which are the same as the research of Yu et al. [25]. Previous studies have reported that the main way to increase the content of toxic metals such as Cd, Cu, Hg, and Zn in agricultural soils in Zhejiang Province is through crop harvesting and leaching, accounting for 74.43–83.62% of the total output [26]. Cd is the priority control pollutant in agricultural soil in this study, which is the same as that obtained by Kim et al. [27]. They found that Cd pollution from industrial activities is serious or extremely serious to agricultural soil.

According to the survey, before 1990, pigs and poultry were fed coarse grains and green feed without any commercial feed, so there was no chance of toxic metal pollution. After 1990, the number of large-scale pig farms in South Lake, Jiashan, and Tongxiang increased gradually. A small amount of toxic metals can be added to the feed, which can promote the growth and development of pigs. Therefore, the fermentation of pig manure containing toxic metals as organic waste materials directly applied to agricultural fields resulted in a significant increase of toxic metals in agricultural soils. According to Wang et al. [28], the organic fertilizer samples contained up to 364–2213 mg·kg^{-1} Cu and 319–4359 mg·kg^{-1} Zn. Through the above analysis, it can be inferred that As pollution is mainly concentrated in these areas, which may be due to the long-term high-intensity use of pig manure, resulting in the serious accumulation of toxic metals in these areas.

3.2.2. Spatial Variation of Toxic Metals Affected by Soil Types and Crop Cultivation Methods

Data obtained from 250 samples were subjected to statistical analysis using the risk screening value of the National Soil Environmental Quality Standard as the limit to calculate the arithmetic mean, single-factor pollution index (Pi), and comprehensive pollution index (Pcom) of the samples, as shown in Table 3. The soil types ranged from blue–purple clay distributed in the lake group, disc edge, and low ground in the mainland, and silt-clayey yellow mottled, blue clayey, and green powder paddy soils on the central water network plains to the semi-sandy soil and silt-loamy paddy soil on the coastal (along with the river plains); the comprehensive pollution indices obtained for these soils were 0.614, 0.697, 0.602, 0.584, 0.626, 0.632, 0.464, and 0.495, respectively. Toxic metal pollution shows a clear spatial variation, decreasing from the mainland to the coastal plain. In addition, research has demonstrated that transportation plays a significant role as a conduit for heavy metals to move from land to rivers, closely correlating with distance, freight volume, and traffic emissions [29]. Soil texture also transitions from sticky clay loam soil to light loam soil (except for some silt-loamy paddy soil with a light texture). The gradual decline of the comprehensive pollution index (i.e., the decrease of the degree of soil pollution) is, therefore, related to the type and texture of the soil as well as the terrain (water flow). More toxic metals may accumulate in the soil on low-lying land.

Table 3 reflects the content status and quality evaluation of each element in different soil types. It can be seen that the content of toxic elements and the rate of exceeding the standard in blue clayey paddy soil are relatively high. Taking the background value of soil and the second level limit of soil environmental quality as the standard calculation, the comprehensive pollution indices obtained reached 1.78 and 0.78, respectively. Secondly, the relatively heavy pollution is green powder paddy soil and yellow sand moisture soil, while the element content of silt-loamy paddy soil and semi-sandy soil is close to the background value, basically no pollution. The main reason for this phenomenon may be that the pollutant-carrying capacity of different terrain parts of different soils is different; it may also be caused by the difference in element content in the parent material developed by each soil itself, or it may be related to the physical and chemical properties of the soil. For example, the blue clayey paddy soil is sticky, and the amount of organic matter and cation exchange is the highest of all the soils surveyed, which is conducive to its adsorption and fixation of more foreign elements. While silt-loamy paddy soil and semi-sandy soil are the most sandy of all soils, with relatively low organic matter content, the ability to adsorb and fix foreign elements is very weak, and foreign elements are not easy to accumulate in the soil.

The effects of different crop cultivation methods on the contents of different toxic metals in soils were significantly different. It can be seen from Figure 6 that the content in the soils increased significantly compared to the background values 30 years ago, regardless of the crop cultivation method. Overall, As, Cd, Cr, Cu, Hg, Ni, Pb, and Zn were most abundant in soils grown from grain and oil crops. Toxic metals are least abundant in fruit tree-grown soils. However, taking the screening values in soil environmental quality as the standard, the exceedances of Cd, Cu, Ni, and Hg in vegetable-grown soils were significantly higher than those in soils grown in grain and oil crops and fruit trees.

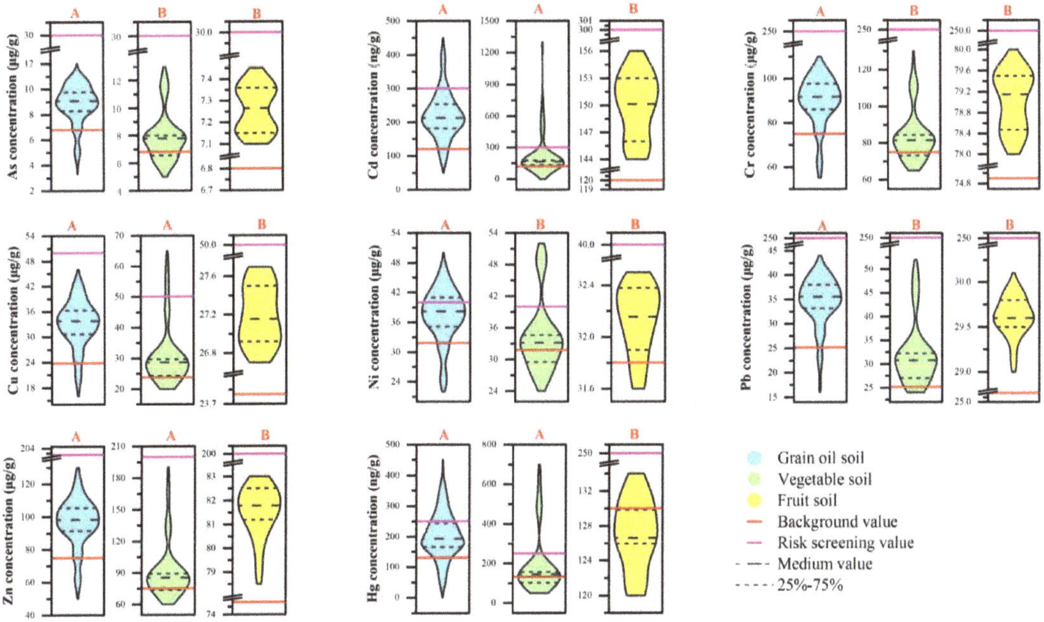

Figure 6. Toxic metal content of soils with different utilization. Red uppercase letters indicate significant differences in heavy metal concentrations among different soil types under various land uses ($p < 0.05$).

4. Conclusions

This study investigated the temporal and spatial variations and mechanisms of toxic metal (As, Cd, Cr, Cu, Hg, Ni, Pb, and Zn) contamination in 250 agricultural soil samples from Jiaxing City. While the levels of all eight toxic metals were found to be within safe limits compared to national risk screening values, there has been a gradual deterioration in toxic metal pollution in the investigated area since the early 1990s, accumulating significantly in a composite pollution manner. Particularly alarming was the contamination of Cd and Hg, which exceeded background levels by more than 1.5 times, potentially elevating environmental and health risks. Moreover, Jiashan County, Nanhu District, and Tongxiang City exhibited the most severe comprehensive pollution, with pollution indices reaching 2.18, 2.02, and 1.97, respectively. Industrialization and urbanization were identified as the primary drivers of temporal variations in toxic metal pollution, while soil types and crop cultivation methods were found to dominate spatial variations in toxic metal pollution. Future research should prioritize investigating the pollution characteristics of toxic metals in crops, thereby enabling the formulation of more effective governance and remediation strategies.

Author Contributions: M.C.: Writing—original draft. Y.J.: writing—review and editing. X.L.: investigation. J.H.: software. Y.Y.: methodology. L.H.: data curation. W.W., W.Z., and F.Z.: methodology. C.H.: methodology, investigation, resources, writing—review and editing. All authors have read and agreed to the published version of the manuscript.

Funding: This research was funded by the Zhejiang Province Key R&D project (No. 2021C03025; 2017C03008; 2015C03020).

Institutional Review Board Statement: Not applicable.

Informed Consent Statement: Informed consent was obtained from all subjects involved in the study.

Data Availability Statement: The data used to support the findings of this study are included within the article. Some or all the data or models that support the findings of this study are available from the corresponding author upon reasonable request.

Acknowledgments: The authors are highly thankful to the members of our group for their valuable contributions to this paper.

Conflicts of Interest: The authors declare no conflicts of interest. The funders had no role in the design of the study, in the collection, analysis, or interpretation of data, in the writing of the manuscript, or in the decision to publish the results.

References

1. Mounissamy, V.C.; Parihar, R.S.; Dwivedi, A.K.; Saha, J.K.; Rajendiran, S.; Lakaria, B.L.; Patra, A.K. Effects of Co-composting of Municipal Solid Waste and Pigeon Pea Biochar on Heavy Metal Mobility in Soil and Translocation to Leafy Vegetable Spinach. *Bull. Environ. Contam. Toxicol.* **2021**, *106*, 536–544. [CrossRef] [PubMed]
2. Pan, S.; Wang, K.; Wang, L.; Wang, Z.; Han, Y. Risk Assessment System Based on WebGIS for Heavy Metal Pollution in Farmland Soils in China. *Sustainability* **2017**, *9*, 1846. [CrossRef]
3. Jayakumar, M.; Surendran, U.; Raja, P.; Kumar, A.; Senapathi, V. A review of heavy metals accumulation pathways, sources and management in soils. *Arab. J. Geosci.* **2021**, *14*, 2156. [CrossRef]
4. Liu, T.; Wang, Z. Contamination and health risk assessment of heavy metals in soil surrounding an electroplating factory in JiaXing, China. *Sci. Rep.* **2024**, *14*, 4097. [CrossRef] [PubMed]
5. Liu, T.; Ni, S.Y.; Wang, Z. Contamination and health risk assessment of heavy metals in soil surrounding an automobile industry factory in Jiaxing, China. *Front. Environ. Sci.* **2024**, *12*, 1362366. [CrossRef]
6. Liu, T.-T.; Hu, W.-Z.; Wang, G.-F.; Xue, J.-P. Assessment of heavy metal pollution in soil around a plastic metal factory in Jiaxing, China. *IOP Conf. Ser. Earth Environ. Sci.* **2020**, *446*, 032040. [CrossRef]
7. Hu, B.; Zhou, Y.; Jiang, Y.; Ji, W.; Fu, Z.; Shao, S.; Li, S.; Huang, M.; Zhou, L.; Shi, Z. Spatio-temporal variation and source changes of potentially toxic elements in soil on a typical plain of the Yangtze River Delta, China (2002–2012). *J. Environ. Manag.* **2020**, *271*, 110943. [CrossRef] [PubMed]
8. Li, A.; Qian, S.; Li, J. Soil environmental quality and its assessment of rice production for non-environmental pollution in Nanhu district of Jiaxing in Zhejiang Province. *J. Zhejiang Univ. (Agric. Life Sci.)* **2010**, *36*, 74–77.
9. Zhao, G.; Ma, Y.; Liu, Y.; Cheng, J.; Wang, X. Source analysis and ecological risk assessment of heavy metals in farmland soils around heavy metal industry in Anxin County. *Sci. Rep.* **2022**, *12*, 10562. [CrossRef]
10. Lutfullah; Sharma, S.; Rahman, N.; Azmi, S.N.H. Determination of thorium(IV) with rifampicin in synthetic mixture and soil samples by spectrophotometry. *Arab. J. Chem.* **2016**, *9*, S1163–S1169. [CrossRef]
11. Boschetti, W.; Orlando, M.; Dullius, M.; Dessuy, M.B.; Vale, M.G.R.; Welz, B.; de Andrade, J.B. Sequential and simultaneous determination of four elements in soil samples using high-resolution continuum source graphite furnace atomic and molecular absorption spectrometry. *J. Anal. At. Spectrom.* **2016**, *31*, 1269–1277. [CrossRef]
12. Chen, G.; Thompson, A.; Gorski, C.A. Disentangling the size-dependent redox reactivity of iron oxides using thermodynamic relationships. *Proc. Natl. Acad. Sci. USA* **2022**, *119*, e2204673119. [CrossRef] [PubMed]
13. Mensik, L.; Hlisnikovsky, L.; Pospisilova, L.; Kunzova, E. The effect of application of organic manures and mineral fertilizers on the state of soil organic matter and nutrients in the long-term field experiment. *J. Soils Sediments* **2018**, *18*, 2813–2822. [CrossRef]
14. Zhao, R.; Guan, Q.; Luo, H.; Lin, J.; Yang, L.; Wang, F.; Pan, N.; Yang, Y. Fuzzy synthetic evaluation and health risk assessment quantification of heavy metals in Zhangye agricultural soil from the perspective of sources. *Sci. Total Environ.* **2019**, *697*, 134126. [CrossRef] [PubMed]
15. Zhang, Y.; Song, B.; Zhou, Z. Pollution assessment and source apportionment of heavy metals in soil from lead – Zinc mining areas of south China. *J. Environ. Chem. Eng.* **2023**, *11*, 109320. [CrossRef]
16. Huo, A.; Wang, X.; Zhao, Z.; Yang, L.; Zhong, F.; Zheng, C.; Gao, N. Risk Assessment of Heavy Metal Pollution in Farmland Soils at the Northern Foot of the Qinling Mountains, China. *Int. J. Environ. Res. Public Health* **2022**, *19*, 14962. [CrossRef] [PubMed]
17. Ediagbonya, T.F.; Ajayi, S. Risk assessment and elemental quantification of anthropogenic activities in soil. *Environ. Geochem. Health* **2021**, *43*, 4891–4904. [CrossRef] [PubMed]
18. Dong, X.; Chu, Y.; Tong, Z.; Sun, M.; Meng, D.; Yi, X.; Gao, T.; Wang, M.; Duan, J. Mechanisms of adsorption and functionalization of biochar for pesticides: A review. *Ecotoxicol. Environ. Saf.* **2024**, *272*, 116019. [CrossRef] [PubMed]
19. Xu, Y.; Shi, H.; Fei, Y.; Wang, C.; Mo, L.; Shu, M. Identification of Soil Heavy Metal Sources in a Large-Scale Area Affected by Industry. *Sustainability* **2021**, *13*, 511. [CrossRef]

20. Keshavarzi, A.; Kumar, V.; Ertunc, G.; Brevik, E.C. Ecological risk assessment and source apportionment of heavy metals contamination: An appraisal based on the Tellus soil survey. *Environ. Geochem. Health* **2021**, *43*, 2121–2142. [CrossRef]
21. Cao, J.; Xie, C.; Hou, Z. Ecological evaluation of heavy metal pollution in the soil of Pb-Zn mines. *Ecotoxicology* **2022**, *31*, 259–270. [CrossRef] [PubMed]
22. Chaganti, V.N.; Ganjegunte, G.; Niu, G.; Ulery, A.; Flynn, R.; Enciso, J.M.; Meki, M.N.; Kiniry, J.R. Effects of treated urban wastewater irrigation on bioenergy sorghum and soil quality. *Agric. Water Manag.* **2020**, *228*, 105894. [CrossRef]
23. Li, X.; Liu, L.; Wang, Y.; Luo, G.; Chen, X.; Yang, X.; Hall, M.H.; Guo, R.; Wang, H.; Cui, J.; et al. Heavy metal contamination of urban soil in an old industrial city (Shenyang) in Northeast China. *Geoderma* **2013**, *192*, 50–58. [CrossRef]
24. Yuan, X.; Xue, N.; Han, Z. A meta-analysis of heavy metals pollution in farmland and urban soils in China over the past 20 years. *J. Environ. Sci.* **2021**, *101*, 217–226. [CrossRef]
25. Yu, J.; Li, L.; Yu, L.; Su, Y.; Ma, Y. Analysis of rural soil environmental quality in Jiaxing in 2020. *J. Green Sci. Technol.* **2021**, *23*, 149–152. [CrossRef]
26. Shi, T.; Ma, J.; Wu, F.; Ju, T.; Gong, Y.; Zhang, Y.; Wu, X.; Hou, H.; Zhao, L.; Shi, H. Mass balance-based inventory of heavy metals inputs to and outputs from agricultural soils in Zhejiang Province, China. *Sci. Total Environ.* **2018**, *649*, 1269–1280. [CrossRef] [PubMed]
27. Kim, H.; Lee, M.; Lee, J.-H.; Kim, K.-H.; Owens, G.; Kim, K.-R. Distribution and extent of heavy metal(loid) contamination in agricultural soils as affected by industrial activity. *Appl. Biol. Chem.* **2020**, *63*, 31. [CrossRef]
28. Wang, Y.; Lv, M.; Han, X. Content Analysis of Heavy Metals in Manure Samples from Different Scales of Pig Farms in Ningbo City. *J. Domest. Anim. Ecol.* **2016**, *37*, 55–58+64.
29. Yan, W.; Liu, D.; Peng, D.; Mahmood, Q.; Chen, T.; Wang, Y.; Li, S.; Chen, J. Spatial distribution and risk assessment of heavy metals in the farmland along mineral product transportation routes in Zhejiang, China. *Soil Use Manag.* **2016**, *32*, 338–349. [CrossRef]

toxics

MDPI

Article

Evaluating the Response of the Soil Bacterial Community and Lettuce Growth in a Fluorine and Cadmium Co-Contaminated Yellow Soil

Mei Wang [1,2,*], Xiangxiang Chen [1], Yasir Hamid [2] and Xiaoe Yang [2,*]

[1] School of China Alcoholic Drinks, Luzhou Vocational and Technical College, Luzhou 646000, China
[2] Key Laboratory of Environment Remediation and Ecological Health, Ministry of Education, College of Environmental and Resource Sciences, Zhejiang University, Hangzhou 310058, China
[*] Correspondence: lzywangmei@163.com (M.W.); xeyang@zju.edu.cn (X.Y.); Tel.: +86-13551108496 (M.W.); +86-13858085377 (X.Y.)

Abstract: The impact of cadmium (Cd) and fluorine (F) on plant and human health has provoked significant public concern; however, their combined effects on plant and soil bacterial communities have yet to be determined. Here, a pot experiment was conducted to evaluate the effects of exogenous F, Cd, and their combination (FCd) on lettuce growth and soil bacterial communities. The results revealed that F and Cd concentrations in lettuce ranged from 63.69 to 219.45 mg kg^{-1} and 1.85 to 33.08 mg kg^{-1}, respectively, presenting lower values in shoots than in the roots. Moreover, low contamination levels had no discernable influence on lettuce growth, but showed a synergistic negative on plant biomass when exogenous F and Cd exceeds 300 and 1.0 mg kg^{-1}, respectively. The results of 16S rRNA gene sequencing indicated that the most abundant bacterial community at the phylum level was *Proteobacteria*, with the relative abundance ranging from 33.42% to 44.10% across all the treatments. The contaminants had little effect on bacterial richness but impacted the structure of bacterial communities. The PCoA showed that compartment and contaminants were the primary contributors to the largest source of community variation, while the VPA indicated that F and Cd synergistically affected the bacterial communities. In turn, lettuce plants could enhance the resistance to the combined stress by increasing the relative abundance of *Oxyphotobacteria*, *Subgroup 6*, *Thermoleophilia*, and *TK10* classes in the rhizosphere.

Keywords: cadmium; fluorine; chlorophyll; diversity; structure; synergistic effect

1. Introduction

Heavy metals (HMs) contamination is a pressing environmental challenge globally that presents a significant threat to human health [1,2]. Among all HMs, cadmium (Cd) is a prominent pollutant, found in several regions in China [3,4]. Fluorine (F) is the 13th most abundant element in the earth's crust and is widely recognized for its adverse effects on health in many regions around the globe [5,6]. F is often found in conjunction with Cd in the environment [7,8]. Previously, Cd was reported as a hidden toxin in an endemic fluorosis area [9]. Further, F-contaminated soil was also found in a high geochemical Cd area [10]. In addition, phosphorus fertilizer also contributes as a major source of Cd and F inputs into farmlands, and contains high concentrations of these elements, potentially increasing the concentration of Cd and F in soils and plants [6,11].

Plants exposed to F or Cd stress exhibit several negative impacts on physiological, biochemical, and morphological parameters, such as enzyme activities and metabolic processes [12,13]. Prior studies have documented the accumulation, interaction, and fractionation of Cd and F in soil-crop systems [7,14]. However, the effect on plants varied with the plant species. For instance, Chen et al. [7] reported that co-contamination of F and Cd has a synergistic negative effect on plant growth, and the contaminants are mainly accumulated in the leaves of radish (*Raphanus sativus* L.). However, it was discovered that Cd addition may restrict the F accumulation in oilseed rape (*Brassica napus* L.), while F treatment promoted the Cd accumulation in the tissues [15]. Therefore, it is vital to investigate the co-effect of F and Cd on more crop species that are closely related to human health.

Microorganisms are often more susceptible to contaminants than plants that inhabit the same soil [1,16]. Previously, numerous researches determined the effects of Cd alone or in combination with other HMs on plant growth and bacterial communities. For example, it was reported that HMs pollution has a considerable impact on the diversity, composition, and distribution of bacterial communities in soil [17]. Further, the effect of F on soil microorganisms in the soil-plant system was observed. It was found that the relative abundance of *Proteobacteria* phylum decreased with an increase in F spiking (0, 600, 1000 mg kg^{-1}) in soils [18], where the F addition with 1000 mg kg^{-1} changed the bacterial diversity [19]. Moreover, the co-effects of HMs and rare earth elements (REEs), as well as the combined effects of F and Cl on soil microorganisms have garnered increasing interest. Luo et al. [16] reported that both REEs and HMs regulated the structure and composition of bacterial communities, whereas their combined impacts were higher than the individual effects of HMs or REEs. According to reports, the combination of pyrene and Cd had a greater biocidal influence on soil microorganisms than either Cd or pyrene alone [20], and the interaction between polycyclic aromatic hydrocarbons (PAHs) and HMs altered the fate of pollutants and microbial communities in soils [21]. However, to the best of our knowledge, studies on the effects of combined F and Cd on soil microorganisms are very scarce.

Lettuce (*Lactuca sativa* L.) is a globally cultivated vegetable and is commonly used for standard toxicity tests of harmful elements [22]. Previous studies have investigated the toxicity, accumulation, and translocation of co-existing metal elements in lettuce, such as Cd-Zn [23]. However, to date, the effects of combined F and Cd on lettuce plants, which may raise a potential risk to human health through food chains, have never been investigated. Considering the complex interactions of F and Cd on microorganisms and plants, the present work investigated the effects of soil spiked Cd and F on lettuce growth, soil bacterial communities, and the interactions among the pollutants, plant, and soil bacteria using a pot experiment. The main objectives of the present study were to: (1) investigate the effects of Cd or/and F on lettuce growth, and (2) explore the response of soil bacterial communities to spiked Cd or/and F in the soil-plant system. We hypothesized that: (1) there were hormetic dose—response relationships for the effect of Cd and F on lettuce growth; (2) F and Cd have a synergistic effect on plant growth and soil bacterial communities.

2. Materials and Methods

2.1. Soil Aging

The surface soil (0–20 cm) was collected from farmland in the southwest of China (105°19′26″ E, 27°22′48″ N). The basic properties of soil and analysis methods are listed in Table S1. The soil was classified as yellow soil based on its properties and the Chinese soil classification system. Before the pot experiment, the collected soil was air-dried and sieved to <2 mm. The exogenous concentrations of Cd and F in soil (Table 1) were designed according to the data of field investigation and information from previously published literature [7,15]. The soil was spiked with the appropriate amount of Cd [$Cd(NO_3)_2 \cdot 4H_2O$] or/and F (NaF), by adding an equal volume (1.25 L) and thoroughly mixed into the soil with a total weight of 7.5 kg for each treatment. Then, the soil samples were incubated for three months at a moisture level of 60% water-holding capacity to achieve chemical equilibration in the greenhouse with the conditions were 20–25 °C, relative humidity 60–70%, and light intensity 180 $\mu Mm^{-2}\ s^{-1}$ during a 16/8 h day/night photoperiod. At the end of the aging period, 1.25 kg of soil was transferred into each plastic pot of 15 cm diameter and 12 cm height.

Table 1. Exogenous concentrations of F and Cd in soil. Treatment included with F (F1 to F5), Cd (Cd1 to Cd5), and FCd (FCd1 to FCd5). A soil without addition was conducted as control treatment (CK).

Treatment	F mg kg^{-1}	Cd mg kg^{-1}	Treatment	F mg kg^{-1}	Cd mg kg^{-1}	Treatment	F mg kg^{-1}	Cd mg kg^{-1}
CK	0	0						
F1	50	0	Cd1	0	0.3	FCd1	50	0.3
F2	100	0	Cd2	0	0.6	FCd2	100	0.6
F3	300	0	Cd3	0	1.0	FCd3	300	1.0
F4	500	0	Cd4	0	2.0	FCd4	500	2.0
F5	1000	0	Cd5	0	5.0	FCd5	1000	5.0

2.2. Plant Culture

Plant culture was carried out in a greenhouse at Zhejiang University. Seeds of Frisee lettuce (Huayu, Hebei, China) were bought from the local market. 20 uniform and plump seeds were evenly sown in each plastic pot. Each treatment consists of six replicates which were carried out under a completely randomized block design. The seedlings were thinned to one lettuce plant left in each plot on the 7th day of growth. The moisture content in the pots was kept at 60% water-holding capacity during the whole experimental period. The seedlings were harvested after three months of growth. Due to the limited plants and quantity of rhizosphere soils, samples collected from two replicates were combined for further analysis.

2.3. Plant and Soil Analysis

Chlorophyll was extracted with a mixture of acetone and ethyl alcohol and analyzed using a microplate reader (Epoch 2, BioTek, Thermo Fisher Scientific, Waltham, MA, USA) [24]. Plants were separated into roots and shoots, washed with tap water and deionized water, and then dried. The fresh weight (FW) of shoots was gauged by a balance with an accuracy of 0.001 g, and the dry weight (DW) was measured after drying at 65 °C for 72 h, while the biomass of roots remained unmeasured as certain parts of the roots remained in soils. The rhizosphere and bulk soils were collected by following the method developed by Edwards et al. [25]. Soil collected from two pots was mixed and later divided into two parts, where one was stored at −80 °C for soil DNA extraction, and the remainder was air-dried to analyze physical and chemical properties. All materials and reagents for extracting and analyzing soil microorganisms were sterilized before use.

The dried plant samples were ground and sieved to <0.43 mm, and the air-dried soil samples were passed through 2 mm and 0.15 mm sieves for analysis. The total Cd content in soil and plant was analyzed with inductively coupled plasma-mass spectrometer (ICP-MS, Agilent 7500a, Agilent Technologies Co., Ltd., Santa Clara, CA, USA) with mixed acids digestion method [24]. The analysis of F was performed using an ion-sensitive electrode method with an electrode (PF-1-01, Shanghai INESA, Shanghai, China) after samples were digested with NaOH [26]. The quality control of the measurements was performed using standard soil (GBW07405), wheat (GBW100495), and tea (GBW10016) samples bought from the Institute of Geophysical and Geochemical Exploration, China. Translocation factor (TrF) was calculated to evaluate the mobility of Cd or F from the roots to the shoots of lettuce. The bio-concentration factor (BCF) was used to assess the ability to transport and accumulate F and Cd in the lettuce plant. The equations are listed as below:

$$TrF = \frac{concentration\ of\ F\ or\ Cd\ in\ shoots}{concentration\ of\ F\ or\ Cd\ in\ roots} \tag{1}$$

$$BCF = \frac{concentration\ of\ F\ or\ Cd\ in\ roots}{concentration\ of\ available\ F\ or\ Cd\ in\ soils} \tag{2}$$

2.4. DNA Extraction, PCR Amplification, and 16S rRNA Gene Sequencing

Microbial DNA in soil was extracted by PowerSoil® DNA Isolation Kit (MOBIO, Carlsbad, CA, USA), and quantified with Nanodrop spectrophotometer (GENE, Shanghai, China). The V3–V4 region of the bacterial 16S rRNA gene was amplified by the primer sets 338F (5′-ACTCCTACGGGAGGCAGCA-3′) and 806R (5′-GGACTACHVGG GTWTCTAAT-3′) with barcode [27]. PCR amplifications were first carried out using KOD FX Neo (TOYOBO, Japan). Briefly, the PCR reactions were amplified by thermocycling (25 cycles at 95 °C for 30 s, 50 °C for 30 s, and 72 °C for 40 s) after initialization for 5 min at 95 °C, followed by a 7 min final elongation at 72 °C. The PCR products were purified through VAHTSTM DNA Clean Beads (VAZYME, Nanjing, China). After that, the second round of PCR was performed in a 20 μL reaction which contained 10 μL 2 × Phμsion HF MM, 2.5 μL of each primer (2 μM), and 5 μL purified PCR products. Thermal cycling conditions were as follows: an initial denaturation at 98 °C for 30 s; followed by 10 cycles at 98 °C for 10 s, 65 °C for 30 s, and 72 °C for 30 s, with a final extension at 72 °C for 5 min. The PCR products was certified on 1.8% (m/v) agarose gels and quantified by the ImageJ v1.53 g software. Moreover, samples were mixed in equidensity ratios and purified with an e.Z.N.A.TM Cycle-Pure Kit (OMEGA, Connecticut, USA). The products were certified by agarose gel electrophoresis, followed by the recovery of the target fragment by using the Monarch DNA Gel Extraction Kit (GENE, Shanghai, China). Sequencing libraries were validated using a Qsep-400 Multi-Channel Bio-Fragment Analyzer (BIOPTIC, Jiangsu, China). Finally, sequencing was performed using an Illumina HiSeq 2500 platform (Illumina Inc., San Diego, CA, USA). The generated sequences were submitted in the Sequence Read Archive (SRA) database of NCBI under accession number PRJNA1101703.

2.5. Processing of Sequencing Data

The raw sequencing reads were primarily filtered using Trimmomatic v0.33 [28] based on the window size of 50 bp, and the reads will be cut from the start of the window once the average Q-score within the window is lower than 20. Using Cutadapt v1.9.1 [29] to identify and remove primer sequences with a maximum mismatch accepted of 20% and a minimum coverage of 80%. The paired-end reads were then assembled using Usearch v10.0 [30] followed by chimera removal using Uchime v8.1 [31]. The parameters in Usearch were set as: minimum length of overlap (10 bp), minimum similarity within overlapping regions (90%), and maximum mismatch accepted (5 bp). Chimera removal involves three steps in Uchime: (1) split query sequence into non-overlapped chunks, (2) compare these chunks with the reference database to identify the best hit of each chunk in the database and further define two best parent sequences, and (3) compare the query sequence with the two parent sequences while if a fragment with over 80% similarity to query sequence is found on both parents, this query sequence will be defined as chimera sequence. The generated sequences that had a minimum of 97% similarity were clustered into operational taxonomic units (OTUs) by using the Usearch [30], and OTUs with less than 0.005% of all sequences were removed [32]. Afterward, taxonomic annotation was processed with classify-sklearn in Qiime2 v2020.6 according to the naive Bayes classifier-based method by using SILVA132 as the reference database [33]. The classifier needs to be trained before use to "learn" which features can be used for classification, and the confidence level of the classifier was set to 0.7.

2.6. Statistical Analysis

One-way analysis of variance (ANOVA) based on LSD post hoc test was used to examine the variations in parameters of soil and plants among different treatments in SPSS v22.0 (IBM, Armonk, New York, USA)). Relevant figures were visualized in OriginPro 2017 (OriginLab, Northampton, MA, USA). To determine the α-diversity, the species number and Chao1 index (indicate species richness), as well as Shannon and Faith's PD (indicate species diversity) were calculated by Qiime2. The β-diversity was analyzed in R v.4.1.1. The principal coordinate analysis (PCoA) based on the Weighted Unifrac metric was used to analyze the differences in bacterial community structures. Permutational multivariate analysis of variance (PERMANOVA) was conducted to assess the differences in community composition in R by using Vegan's function adonis [34]. To analyze the correlation between bacterial species and soil properties, the redundancy analysis (RDA) was performed using binary Jaccard metric. Moreover, variance partitioning analysis (VPA) was used to explain the effects of F and Cd on the distribution of soil bacterial communities in R by using Vegan's function varpart. Least discriminant analysis (LDA) based on the Kruskal—Wallis rank sum test ($\alpha = 0.05$) was used to determine the species with significant differences in different treatments [35]. The relationships between the relative abundance of rhizosphere-specific bacterial taxa and parameters of soil and plants were analyzed using Pearson's correlation test, and a heatmap was performed. Phylogenetic investigation of communities by reconstruction of unobserved states 2 (PICRUSt2) was used to predict the metagenome functional content based on the Kyoto Encyclopedia of Genes and Genomes (KEGG) pathway for function annotation [36].

3. Results

3.1. Biomass and Chlorophyll Content of Lettuce

The effect of exogenous F or/and Cd in soil on lettuce growth was visible (Figure S1). The weight (FW and DW) of lettuce shoots showed a decreased tendency with an increasing concentration of spiked pollutants (Figure 1A,B). Compared with CK, pollutant supplements lower than 300 mg F, 1.0 mg Cd, and combined 100 mg F and 0.6 mg Cd kg^{-1} soil significantly declined the DW of lettuce shoots ($p < 0.05$, Figure 1B). Overall, synergistic negative effects of F and Cd on lettuce shoot weight were observed in the soil spiked with combined F ($\geq$300 mg kg^{-1}) and Cd ($\geq$1.0 mg kg^{-1}) compared with that of the single Cd or F treatment (Figure 1A,B).

In the case of sole F treatment, the chlorophyll a, b, and total chlorophyll concentration (mg g^{-1} FW) remained in the range of 2.60–3.17, 0.683–0.838, and 3.29–4.00, respectively (Figure 1C–E). Compared with CK, low F addition (F1) in soil increased chlorophyll concentrations, but higher F had no significant effect on chlorophyll contents (Figure 1C–E). For the Cd and FCd treatments, compared with CK, there was no statistical difference in chlorophyll concentrations with a lower rate of Cd ($\leq$2.0 mg kg^{-1}), but higher Cd addition tends to decrease the chlorophyll contents (Figure 1C–E). The negative synergy of F and Cd was detected with a higher spiking rate of F and Cd (500 and 2.0 mg kg^{-1}), respectively (Figure 1C–E).

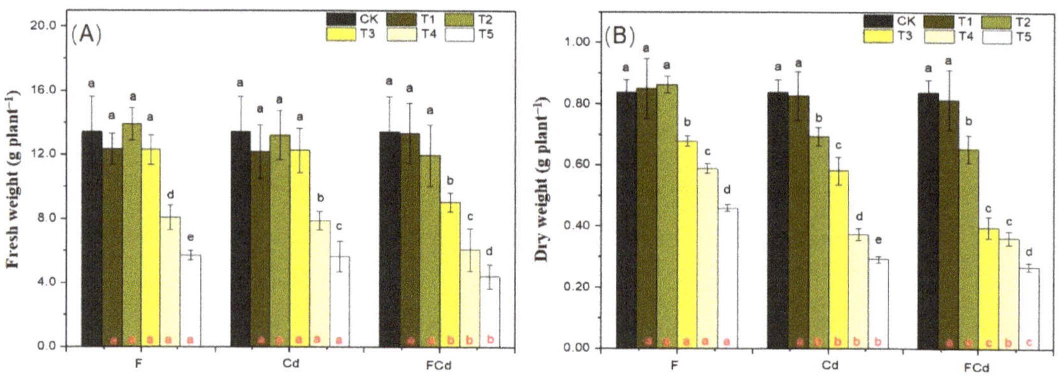

Figure 1. *Cont.*

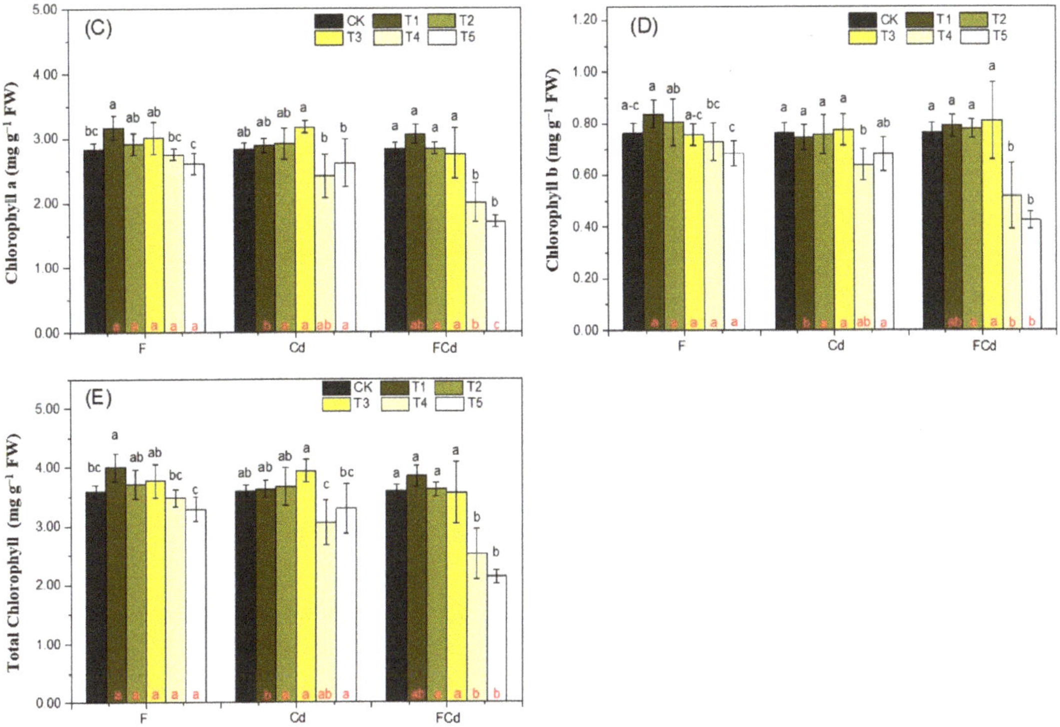

Figure 1. The effect of exogenous F or/and Cd in soil on fresh weight (**A**), dry weight (**B**), chlorophyll a (**C**), chlorophyll b (**D**), and total chlorophyll (**E**) of lettuce. Values are the mean ± SD (n = 3). Different black letters indicate a significant difference at $p < 0.05$ exposure to different levels of the element. Different red letters represent a significant difference at $p < 0.05$ exposure to same concentration of pollutants in Cd, F, and FCd treatments.

3.2. Concentration, Translocation and Bio-Concentration Factor of Cd and F in Lettuce

Concentrations of Cd and F in shoots and roots of lettuce were elevated in Cd and F spiked soils (Figure 2). It was found that F concentration in the shoots (63.69–109.85 mg kg^{-1} DW) was significantly lower than that in the roots (95.47–219.45 mg kg^{-1} DW, Figure 2A). Similar trend was observed in the distribution of Cd in lettuce plants since Cd concentration in shoots and roots was in the range of 1.85–10.67 and 1.97–33.08 mg kg^{-1} DW, respectively (Figure 2B). Meanwhile, the TrF of F (0.499–0.754) first showed an increase and then tended to decrease with an increase in F in the soil (Figure 2C). A declining trend was observed for the BCF of F with values between 4.86 and 31.37 (Figure 2E). Similarly, the BCF of Cd (from 6.84 to 12.96) increased initially and then reduced in response to increasing soil Cd concentration (Figure 2F), but the TrF of Cd (0.313–0.941, except for Cd1 treatment) showed a declining trend (Figure 2D). It was observed that in the FCd treatments, the TrF of F (except for FCd4) was slightly lower than that in single F treatment ($p > 0.5$, Figure 2C); however, the TrF of Cd (except for FCd5) was often higher in Cd treatment than in FCd treatments (Figure 2D).

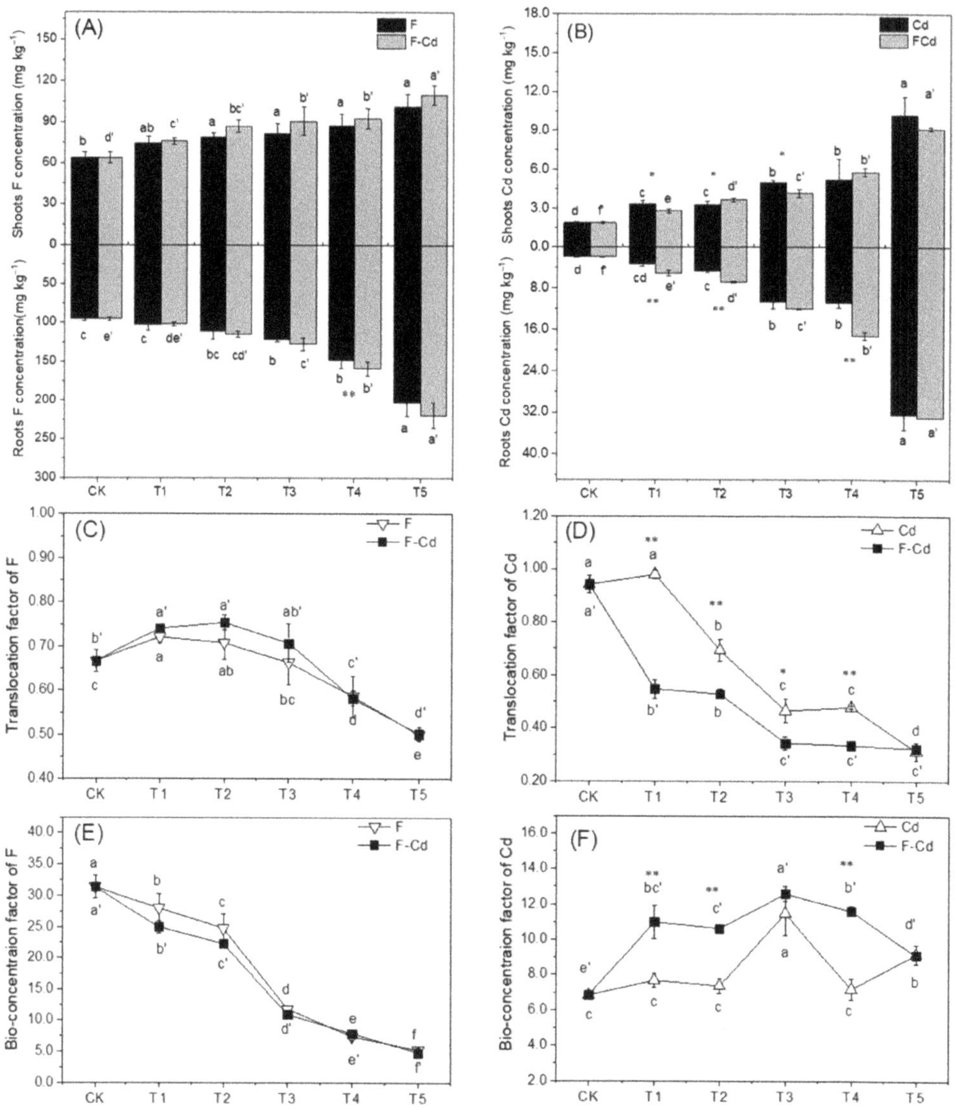

Figure 2. Concentration of F (**A**) and Cd (**B**) in shoots and roots of lettuce, as well as the translocation factor and bio-concentration factor of F (**C**,**E**) and Cd (**D**,**F**) in the plant. Values are the mean ± SD (n = 3). Different letters indicate a significant difference at $p < 0.05$ exposure to different levels of the element. * and ** represent a significant difference at $p < 0.05$ and $p < 0.01$, respectively, when exposure to the same level of pollutants in F, Cd, and FCd treatments.

3.3. Alpha Diversity of Soil Bacteria

Across all samples, 1,893,373 high-quality clean tags were obtained, which were grouped into 1021 OTUs based on the 97% sequence similarity cutoff. The OTUs ranged from 903 to 1018, as observed in the flat rarefaction curves (Figure S2), indicating that all samples had attained saturation with OTUs. The Chao1 index ranged from 952 to 1052 with the highest richness index observed in the rhizosphere soil in FCd treatment (Table 2). The Shannon and Faith's PD index were in the range of 7.59–7.96 and 50.46–57.30, respectively (Table 2). According to the values of OTUs, Chao1, Shannon, and Faith's PD, the rhizosphere had higher species richness and evenness of bacteria than the bulk soil, while no obvious effect was observed for bacterial α-diversity under different pollutants (Table 2).

Table 2. Summary of alpha diversity metrics. Values are the mean ± SD (n = 3). CKB, CdB, FB, and FCdB represent soil samples collected from bulk soil in treatment of CK, Cd4, F4, and FCd4, respectively. CKR, CdR, FR, and FCdR represent soil samples collected from rhizosphere soil in treatment of CK, Cd4, F4, and FCd4, respectively. Different letters represent a significant difference at $p < 0.05$ exposure different treatments. The same as below.

Sample ID	OTUs	Chao1	Shannon	Faith's PD
CKB	915.00 ± 2.65 c	962.82 ± 5.34 c	7.66 ± 0.03 bc	50.97 ± 0.48 c
FB	902.67 ± 4.93 c	952.34 ± 9.00 c	7.59 ± 0.10 c	50.46 ± 0.46 c
CdB	965.67 ± 19.30 b	1001.66 ± 14.17 b	7.79 ± 0.02 abc	54.40 ± 0.31 b
FCdB	967.67 ± 14.15 b	999.09 ± 29.13 b	7.82 ± 0.04 abc	54.15 ± 0.26 b
CKR	1014.00 ± 36.76 a	1040.26 ± 31.67 a	7.73 ± 0.35 abc	56.91 ± 1.86 a
FR	1009.67 ± 3.06 a	1041.43 ± 2.69 a	7.79 ± 0.02 abc	57.36 ± 0.19 a
CdR	974.33 ± 13.50 b	1029.03 ± 6.10 ab	7.90 ± 0.06 ab	54.86 ± 1.70 b
FCdR	1018.00 ± 8.19 a	1051.79 ± 18.44 a	7.96 ± 0.03 a	57.30 ± 0.49 a

3.4. Composition of the Bacterial Community

The dominant ten phyla and genera of soil bacteria were selected to generate histograms of different samples (Figure 3A,B). Bacterial sequences were primarily composed of *Proteobacteria* phylum (ranged from 33.42% to 44.10%, average of 39.45%) among all the samples (Figure 3A). The *Proteobacteria* phylum in the soil mainly consists of *Alpha-* and *Gamma-proteobacteria* with a relative abundance in the range of 17.27–25.89% and 14.85–19.80%, respectively (3B). As shown in Figure 3A, the relative abundances of *Firmicutes*, *Chloroflexi*, *Gemmatimonadetes*, and *Actinobacteria* at the phylum level varied greatly in the bulk and rhizosphere soils. As for the genus level (Figure 3B), the prominent bacteria were *uncultured_bacterium_f_Chitinophagaceae* (4.13–7.24%), *Bryobacter* (3.26–8.32%), and *Sphingomonas* (2.24–4.43%). Meanwhile, the relative abundances of *Bryobacter*, *Gemmatimonas*, *Luteimonas*, *uncultured_bacterium_f_Gemmatimonadaceae*, and *uncultured_bacterium_f_Microscillaceae* at the genus level varied greatly in the bulk and rhizosphere soils (Figure 3B).

As for bacterial community structures (the PCoA result), the first axis (>35.90%) classified the samples by compartments (rhizosphere and bulk soils) while the second axis (>12.61%) was defined by pollutants (F, Cd, and FCd) (Figure 3C,D). This was in line with the results of PERMANOVA, which showed that the compartment (41.5%, $p < 0.01$) and pollutants (26.3%, $p < 0.05$) were the primary and secondary sources of community variation, respectively (Table S2). Additionally, LDA was used to identify biomarkers with statistically significant differences between groups (Figure S3). The results showed that *Alphaproteobacteria* class, *Rhizobiales* order, *Gemmatimonadetes* phylum was the main specific bacterial taxa in lettuce rhizosphere soil under Cd, F, and FCd stress, respectively; while that in the bulk soil was phylum of *Firmicutes*, *Gemmatimonadetes*, and *Proteobacteria*, respectively. These results certified that the supplement of F, Cd, and FCd had distinct effects on the soil bacterial community.

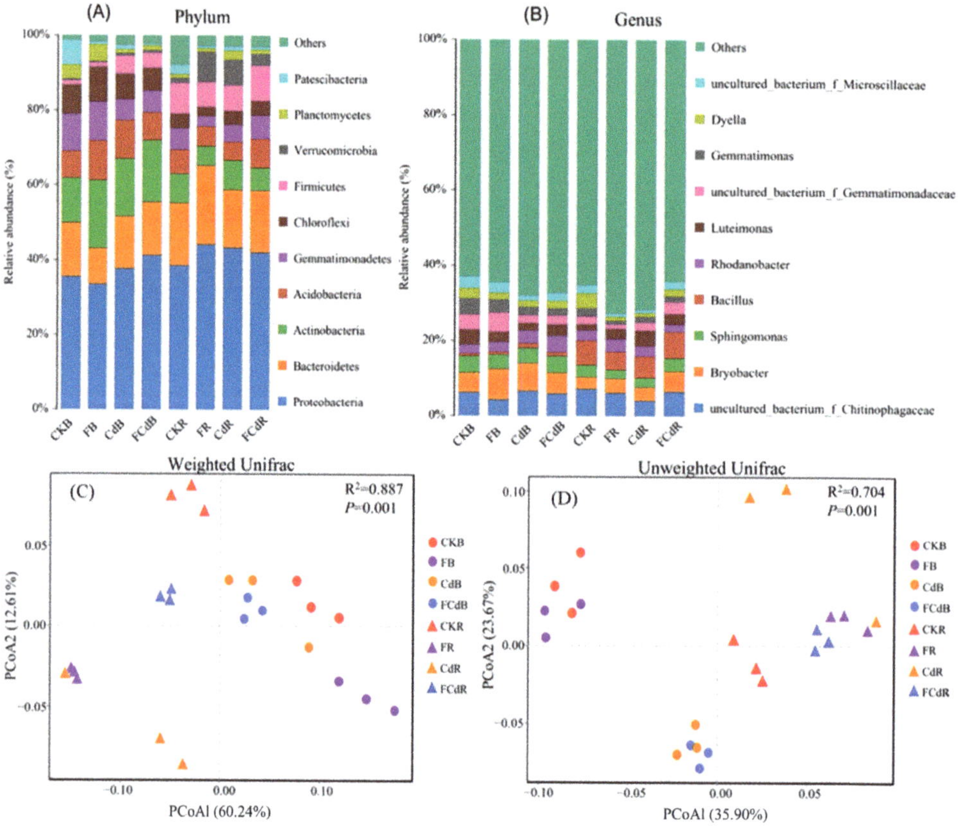

Figure 3. Relative abundance of dominant bacteria at the phylum level (**A**) and genus level (**B**). The principal coordinate analysis (PCoA) plots for visualization based on Weighted Unifrac (**C**) and Unweighted Unifrac (**D**) distances among the bacterial communities.

3.5. Soil Properties and Pollutants Affecting the Bacterial Community

The RDA showed a positive correlation between bacterial communities in the rhizosphere and the soil pH while a negative correlation was observed in bulk soil (Figure 4A). Inversely, bacterial communities in rhizosphere soil exhibited a negative correlation with soil AP and AK but positively correlated with that of bulk soil. In spite of that, F and Cd stresses (TCd, TF, ACd, AF) have no consistent effect on the bacterial community in bulk and rhizosphere soils. Soil F and Cd showed positive effects on the bacterial composition (Figure 4A). This was verified by VPA that the unique contribution of F and Cd to the variation in bacterial communities were 9.60% and 13.68%, respectively, while the pollutants' contribution overlapped by 11.15% (Figure 4B). These results indicate that F and Cd had a synergistic effect on bacterial communities in the soil.

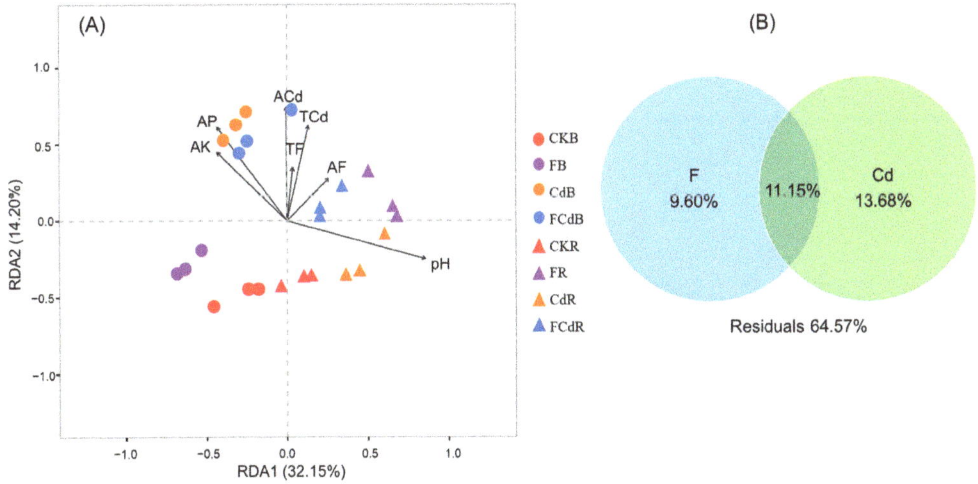

Figure 4. Redundancy analysis (RDA) of soil bacterial communities and selected soil properties (**A**) and variance partitioning analysis (VPA) of soil bacterial communities with soil F and Cd (**B**). AK, AP, ACd, AF represent the available phosphorus, potassium, cadmium, fluorine in soil, respectively. TCd and TF represent the total cadmium and fluorine in soil, and pH means soil pH. The magnitude and direction of correlation are shown by the length and angle of the arrows. F and Cd in the right picture represent the total and available fluorine or cadmium concentration in the soil.

3.6. Response of Bacterial Communities to F and Cd Stresses in Rhizosphere Soil

In comparison with CK, the rhizosphere soil of FCd treatment contained the largest number of unique enriched OTUs (134), while the lowest number of the unique depleted OTUs was 128 as observed in the rhizosphere of F treatment (Figure 5A). Further, the rhizosphere soil treated with Cd owned the least number of unique enriched OTUs (2) and depleted OTUs (5) across all treatments (Figure 5A). Among the three treatments, OTUs belonging to the *Alphaproteobacteria* class were the mainly enriched taxa in rhizosphere soil, followed by the *Verrucomicrobiae* class (Figure 5B). Compared to single F or Cd treatment, the FCd stress decreased the relative abundance of the *Verrucomicrobiae* class but increased that of the *Thermoleophilia* and *Clostridia* classes in the rhizosphere soil (Figure 5B). With regard to the depleted OTUs, there were exclusionary effects on soil bacteria, since 228, 72, and 172 OTUs were significantly depleted in the rhizosphere of F, Cd, and FCd treatment, respectively, corresponding to CK (Figure 5C). The majority of depleted OTUs in the F and FCd treatments are the *Actinobacteria* class, while those in the Cd treatment are the *Gammaproteobacteria* class (Figure 5C). Noteworthy, the FCd supplement increased the relative abundance of *Oxyphotobacteria*, *Subgroup_6*, *Thermoleophilia*, and *TK10* classes corresponding to single F or Cd stress, while that of *Saccharimonadia*, *Chloroflexia Deltaproteobacteria*, and *Gemmatimonadetes* declined in the rhizosphere soil (Figure 5C). These results indicated that the rhizosphere effects of bacteria in lettuce under F, Cd, and FCd stresses owned certain similarities but also showed apparent differences.

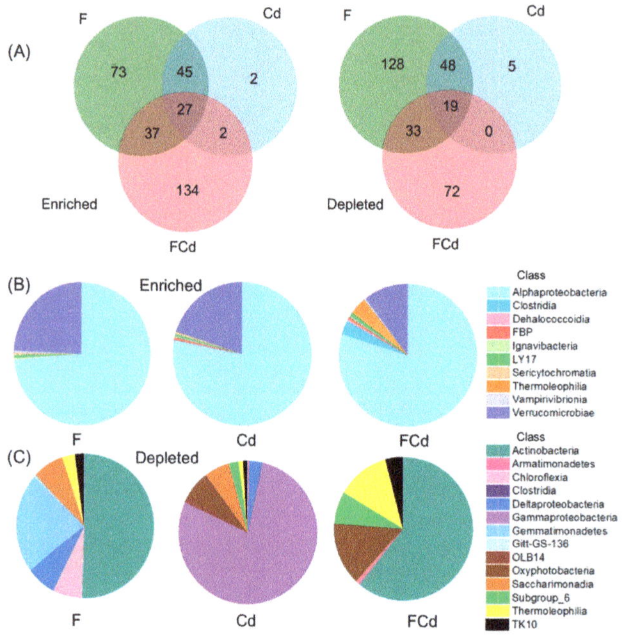

Figure 5. Enriched and depleted of OTUs in rhizosphere soil under different pollutants stress. (**A**) Venn diagrams displayed the enriched and depleted OTUs numbers under F, Cd, and FCd treatments ($p < 0.05$). Pie charts showed significantly enriched (**B**) and depleted (**C**) bacterial species at the family level in the rhizosphere soil under F, Cd, and FCd treatments ($p < 0.05$). Each segment in the chart is colored by the bacterial class of the corresponding taxa and the size of each segment is proportional to the relative abundance of OTUs assigned to the indicated taxa.

3.7. Relations of Rhizosphere-Specific Bacterial Taxa with Soil Properties and Plant Factors

The relative abundance of rhizosphere-specific bacterial taxa was correlated with soil properties and plant factors (Figure 6). The results showed that soil pH was positively correlated with several rhizosphere-specific bacterial taxa under various stress conditions, including the *Rhizobiaceae* family and *Luteimonas* genus ($p < 0.05$). Further, soil Cd concentration positively correlated with the abundance of certain bacterial communities in the rhizosphere under single Cd stress but negatively correlated with single F stress, and vice versa (Figure 6). Moreover, the relative abundance of the *Luteimonas* genus and *Xanthomonadaceae* family positively correlated with soil Cd, pH, and shoot Cd, but they had a negative influence on shoot weight and TrF of Cd (Figure 6). The majority of enriched OTUs belonging to the *Gemmatimonadetes* phylum were negatively correlated with plant TrF of F, but positively correlated with root DW, and BCF of Cd in FCd treatment (Figure 6). These results indicate that soil, plants, and microorganisms are a unified organic combination, that interrelates and influence each other.

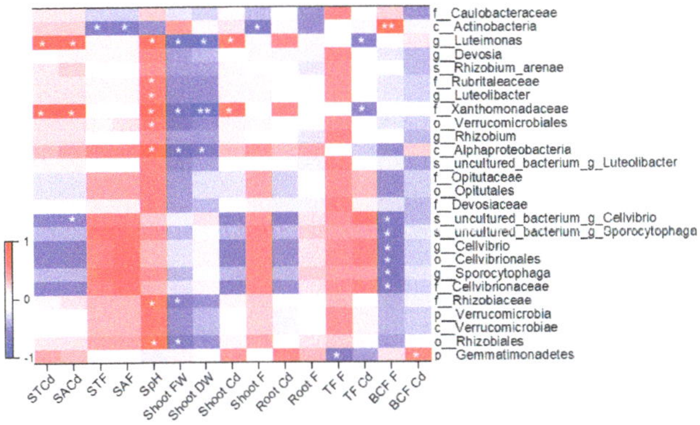

Figure 6. Correlations of the relative abundance of the rhizosphere-specific bacterial taxa of lettuce with soil properties and plant factors. The legend on the left side showed the range of Pearson correlation coefficients (r). The asterisks in the figure indicate significant Pearson's correlations: ** and * represent significant at $p < 0.01$ and $p < 0.05$, respectively.

3.8. Predictive Metagenome Functional Profiling of the Bacterial Community

Exogenous F or/and Cd affected the metabolism of bacteria in both bulk and rhizosphere soils (Figure 7), which was in accordance with the results of PERMANOVA and VPA (Table S2, Figure 4B). The KEGG-based function prediction indicated that the metabolic pathway was most important in soils, with a relative abundance greater than 16% (Figure 7). Further, biosynthesis of secondary metabolites, antibiotics, and amino acids, as well as microbial metabolism in diverse environments and ABC transporters were the main metabolic pathways with a relative abundance above 4% for all treatments (Figure 7). Notably, compared with CK, significant differences in the expression of various genes-related metabolic pathways were observed in soil polluted by F or/and Cd, indicating that bacterial communities altered several metabolic functions for environmental adaptations (Figure 7).

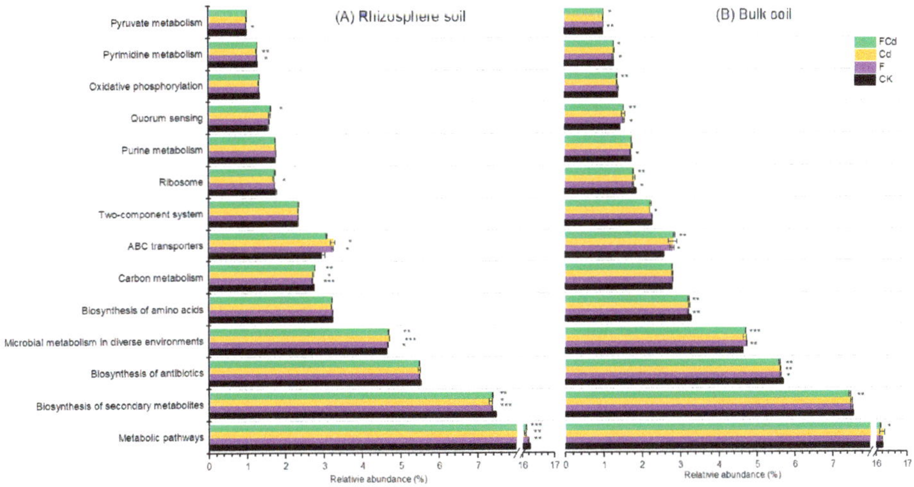

Figure 7. The mainly predictive functional metagenome composition in rhizosphere (**A**) and bulk soils (**B**) of lettuce under F, Cd, and FCd stresses. The *, **, and *** represent statistically significant differences at 0.05, 0.01, and 0.001 level. Only pathways with a mean proportion > 1.0% were displayed (n = 3).

4. Discussion

4.1. Co-Effects of F and Cd on Lettuce Growth

To cope with abiotic stresses such as Cd and F toxicity, plants establish several strategies to mitigate stresses at molecular and physiological levels to keep them healthy in the challenging environment [37–39]. Further, the hormetic dose—response relationships were appropriate for Cd and F in hyperaccumulators, which indicates that low concentrations of pollutants may stimulate plant growth but excessive exposure may lead to toxicity [40]. Previously, lettuce has been reported as a species with a high capacity for Cd accumulation and growth stimulation in the presence of low Cd concentrations [41]. While F is the 13th most abundant and highly electronegative element on the earth, it is abundantly and ubiquitously distributed in the environment [42]. Therefore, low supplementation of F (<300 mg kg^{-1}) or/and Cd (<1.0 mg kg^{-1}) to the soil had no visible effect on lettuce growth as found in the present study (Figure 1).

In spite of that, plants have evolved a well-developed immune system to mitigate abiotic stresses [38,39], it remains difficult to protect lettuce plants from the detrimental effects of excess F or/and Cd stress (Figure 1). It is well established that exposure to F or Cd stress could induce leaf chlorosis thereby reducing plant photosynthesis [37]. In this study, high levels of F or Cd in soil inhibited lettuce growth, and decreased the FW and DW, as well as the chlorophyll contents (Figure 1) which were in line with previous reports [43]. Further, compared with the single F or Cd treatments, the plant weight and chlorophyll contents in FCd treatments (especially in FCd4,5) were decreased which coincides with previous studies on radish and oilseed rape plants [7,15]. In sum, these results confirmed the synergistic effect of F and Cd on the growth of lettuce.

Moreover, the differences in absorption and transportation pathways of Cd and F and their action sites in lettuce plants might explain their negative synergistic effect. The uptake of F from soil into plant roots mainly occurs through a passive diffusion process [44] as well as a partial active process via ion channels, e.g., Cl^- channel [45]. As for Cd, both passive and active transports have been reported for Cd transportation in the plasma membrane of root cells, with the latter requiring the usage of divalent cations, such as Ca^{2+}, Fe^{2+}, and Zn^{2+} [37]. Further, Cd and F ions in solution could co-exist as CdF^+, which could be absorbed by plants [15]. Moreover, F^- can react with Ca^{2+} to generate precipitation of CaF_2 in the solution [12] which could decrease the competition of Ca^{2+} and Cd^{2+} at the surface of roots. These complicated processes consequences in the higher uptake of F and Cd by lettuce roots in FCd treatment than in the single F or Cd treatment. This was consistent with the result of the present study that the concentration and BCF of F and Cd in FCd treatment were repeatedly higher than that in the single F or Cd treatment (Figure 2).

The result in this study was partially similar to Chen et al. [7] that adding Cd into F treatments weakened the F transportation from root to shoot, while adding F into Cd treatments had no discernible effect on the TrF of Cd in radish plants (Figure 2C,D). Although the translocation of F and Cd from root to shoot was unclear, the different translocation properties of the two elements were observed. Apoplastic transport was involved in F transportation [45], and the absorbed F was transported from roots to shoots via the transpiration stream [46]. However, both apoplastic and symplastic pathways were involved in the transport of Cd [37]. It was also reported that a long-distance translocation of Cd may be facilitated by transpiration but can be determined by several factors, such as the amount of Cd sequestered in the vacuole and the availability of other elements [47]. Previously, the translocation of metal—F complexes, such as Al-F, Pb-F, Cd-F, and As-F through the apoplast or via dedicated membrane-bound xenobiotic transporters were reported [47]. In summary, the transport of F and Cd in plants have some independent processes, but the two elements also interact with each other to some extent, and this complicated and changeable system was affected by various ions. As a result, the biochemical and physiological properties of F and Cd as well as the response of the lettuce plant to the stresses induced the synergistic negative effect when F and Cd addition in the yellow soil exceeds 300 and 1.0 mg kg^{-1}, respectively. Considering the concentration of Cd (0.52 mg kg^{-1}) and

F (925 mg kg^{-1}) in the un-spiked soil (Table S1), the total Cd and F concentration in the soil was higher than 1.52 and 1225 mg kg^{-1}, respectively. According to the risk screening values for soil contamination of agricultural land in the State Standard of the People's Republic of China and the evaluation method used in the National soil pollution survey bulletin in China in 2014, the soil was moderately contaminated with Cd. Moreover, F concentration also greatly varies with soil type, thus no specified standard has set the limit value of F in soils in China. However, according to the survey by China National Environmental Monitoring Center, the background value of F in Chinese topsoil was 453 mg kg^{-1} with a range of 191 to 1012 mg kg^{-1}. The soil with an F concentration of higher than 1225 mg kg^{-1} in the present study was polluted. It is worth noting that the high total soil F concentrations were commonly found in endemic fluorosis areas in southwest China [6]. High F and Cd contents in soil not only impair plant growth, but also endanger human health through the food chain and disrupt ecological balance. Moreover, Cd and F in the human body have analogous exposure sources, similar target organs of the toxicities, and analogical clinical symptoms in teeth and bones [48]. Therefore, the migration and transformation of soil Cd and F along the food chain and the health risk assessment of human beings should be directed in future studies.

4.2. Response of the Soil Bacterial Community to F and Cd Stress

It is well established that geographical location, soil types, plant species, cultivation practices, soil compartments, and contaminants pose effects on bacterial communities [25,27]. In this study, the impact of F and Cd on the soil bacterial community was investigated in a soil-plant system. As reported in previous reports [19,27], the rhizosphere and bulk soils were dominated by *Proteobacteria*, *Bacteroidetes*, and *Actinobacteria* (Figure 3). Further, the compartment analysis explained the largest source of variation (Figures 3 and 4; Table S2), indicating the importance of niche filtering that soil microorganisms and lettuce roots may recruit specific microbiota from the bulk soil, which subsequently colonizes in the rhizosphere soil of lettuce. As for the genus level, *Luteimonas* accounted for 1.62–4.14% abundance in this study (Figure 3B) and was positively correlated with the total and available Cd content in the soil (Figure 6). It indicated that *Luteimonas* was Cd tolerant.

Despite the lack of a clear relationship between the bacterial diversity of soil spiked with various pollutants and its bacterial community, it is worth noting that F and Cd were also critical factors and acted synergistically on the soil bacterial community (Figure 4). This was in line with the result that microbial diversity was not significantly affected by HMs contamination, but their composition was significantly affected by the pollution [49]. A similar trend was observed by Luo et al. [16] that HMs and REEs affected the bacterial communities and their combined contributions were greater than the single effects. Organisms can enrich and accumulate F and Cd simultaneously [7,15], as well as high F and Cd concentrations significantly inhibit the microorganisms [19,50]. Therefore, F and Cd probably have a synergistic effect on bacterial communities in soil.

The Rhizosphere acts as an active region for the nutrient and energy exchange in the soil-plant system, and is referred to as a vital barrier for pollutants uptake. Moreover, it is also considered as an important region for pollutant—microorganism interactions [16]. Therefore, the response of bacterial taxa to F and Cd stress, and the interactions of plant-bacteria and soil-bacteria systems were discussed. In our study, compared with the CK, rhizosphere soils spiked with F, Cd, and FCd primarily enriched OTUs belonging to the *Alphaproteobacteria* class (Figure 4B), which might be due to the ecological and biochemical ability to degrade contaminants [51]. Further, OTUs belonging to *Verrucomicrobiae* class were enriched in the polluted rhizosphere soil corresponding to CK (Figure 4B), while *Verrucomicrobia* phylum was enriched in lettuce rhizosphere corresponding to the bulk soil (Figure 3A). Similarly, enrichment of *Verrucomicrobia* in the rhizosphere of maize was reported which might relate to the maintenance of bacterial populations homeostasis in the rhizosphere, since it can form beneficial interactions with plant roots [52]. Furthermore, the enriched OTUs in the rhizosphere under FCd stress were higher than single F or Cd treatment (Figure 4B,C) indicating that the microbial community shifted to strengthen the adaptability of microorganisms towards the pollutants [49]. Interestingly, compared with CK, OTUs belonging to the *Clostridia* class were enriched in the rhizosphere of FCd treatment, but depleted in single F and Cd treatments (Figure 4B,C). *Clostridia* are often described as anaerobic organisms, thus the porosity and permeability of the rhizosphere soil, which is intimately connected to plant roots was improved compared with the bulk soil, especially in yellow clay soil (Table S1). In the *Clostridia* class, multiple plant-pathogenic *Clostridium* species could cause soft rot disease on plants [53], implying that lettuce plants under combined FCd stress have a higher probability of suffering the disease as compared with plants under single F or Cd stress.

Actinobacteria perform a variety of functions in soil-plant systems such as decomposing soil organic matter, elevation in nutrient absorption, disease prevention and suppression in plants, and alleviating plant biotic and abiotic stresses [54]. Compared with CK, OTUs belonging to the *Actinobacteria* class were depleted in the rhizosphere of lettuce in F and FCd treatments but remained unaffected by single Cd stress ($p > 0.05$, Figure 4C), indicating that these taxa were more sensitive to F than Cd stress. However, the *Gammaproteobacteria* class was susceptible to Cd than F stress as these OTUs were significantly depleted in the rhizosphere under Cd stress ($p < 0.05$) with no obvious difference in F and FCd treatments (Figure 4C). Our results were following the findings of Gan et al. [19] who reported no clear relationship between *Gammaproteobacteria* abundance and F content in soil. On the other hand, the synergistic effect of F and Cd on the depleted OTUs belonging to *Saccharimonadia, Chloroflexia, Deltaproteobacteria*, and *Gemmatimonadetes* classes in lettuce rhizosphere (Figure 4C) may contribute to the discrepancy of bacterial communities influenced by F and Cd (Figure 5B). Plants can select specific soil bacterial consortia [55]; therefore, these depleted OTUs which are beneficial for substance circulation in soil and nutrient uptake in plants might participate in inhibiting the growth of lettuce.

Earlier research has confirmed that elevated levels of pollutants in soil might be toxic to a variety of microorganisms due to their effects on metabolic functions [16]. Consistent with a report, bacteria associated with metabolism pathways are the most abundant components in soils [56]. Moreover, in this study, Cd or/and F stresses resulted in a decrease in the relative abundance of genes associated with the degradation and utilization of organic matter (such as carbon metabolism). However, it led to an increase in the genes related to the resistance and transportation of pollutants (such as ABC transporters) (Figure 7). However, due to the limitations of PICRUSt, using multi-omics techniques including metagenomics and transcriptomics to unravel the functional categories responsible for the co-effects of Cd and F in plants is necessary in future research.

5. Conclusions

This study examined the impact of F or/and Cd pollutions in soil on the response of the soil bacterial community and lettuce growth under a pot trial. The results indicated that the concentration of Cd and F in the lettuce plant increased with elevated F and Cd levels in soils, with values ranging from 63.69–219.45 and 1.85–33.08 mg kg^{-1}, respectively. Low concentrations of F and Cd have no obvious effect on plant weight and chlorophyll contents, but excess F and Cd declined these indexes resulting in reduced lettuce growth. Further, soil bacterial diversity was not visibly affected by F or/and Cd; however, the community composition was altered under the stresses. Detailed analysis revealed that compartments were the primary factor contributing to the greatest amount of variation within the community, followed by F and Cd pollution which might have a synergistic effect on the soil bacterial community. In the rhizosphere of lettuce, to alleviate F or/and Cd stresses, bacteria associated with the resistance and transport of pollutants were increased. Future studies should focus on the mechanisms of the synergistic effect of F and Cd on plants and microorganisms, as well as their combined effects on human health by using multi-omics technologies, such as genomics, transcriptomics, proteomics, and molecular biology.

Supplementary Materials: The following supporting information can be downloaded at: https://www.mdpi.com/article/10.3390/toxics12070459/s1, Table S1: Physicochemical properties of the tested soil; Table S2: Determining the environmental variables affecting the microbial community composition using PERMANOVA; Figure S1: Photo of lettuce (*Lactuca sativa* L.) at harvest; Figure S2: The rarefaction curves of samples; Figure S3: Line discriminant analysis (LDA) value distribution histogram for bacterial taxa in rhizosphere soil and bulk soil of lettuce.

Author Contributions: Conceptualization, M.W. and X.Y.; methodology, M.W.; software, X.C.; validation, X.C. and Y.H.; formal analysis, M.W. and X.C.; investigation, M.W.; resources, X.Y.; data curation, X.C.; writing—original draft preparation, M.W.; writing—review and editing, Y.H. and X.Y.; visualization, X.C.; supervision, Y.H.; project administration, M.W.; funding acquisition, M.W. and X.Y. All authors have read and agreed to the published version of the manuscript.

Funding: This research was funded by the research project of Luzhou Vocational and Technical College in 2024 (#LZZX-D-01), and the Fundamental Research Funds for the Central Universities of China.

Institutional Review Board Statement: Not applicable.

Informed Consent Statement: Not applicable.

Data Availability Statement: The original contributions presented in this study are included in this article, and further inquiries can be directed to the corresponding authors.

Conflicts of Interest: The authors declare no conflicts of interest.

References

1. Kushwaha, P.; Neilson, J.W.; Maier, R.M.; Babst-Kostecka, A. Soil microbial community and abiotic soil properties influence Zn and Cd hyperaccumulation differently in *Arabidopsis halleri*. *Sci. Total. Environ.* **2022**, *803*, 150006. [CrossRef] [PubMed]
2. Mai, X.; Tang, J.; Tang, J.; Zhu, X.; Yang, Z.; Liu, X.; Zhuang, X.; Feng, G.; Tang, L. Research progress on the environmental risk assessment and remediation technologies of heavy metal pollution in agricultural soil. *J. Environ. Sci.* **2024**, *149*, 1–20. [CrossRef]
3. Hamid, Y.; Tang, L.; Hussain, B.; Usman, M.; Lin, Q.; Rashid, M.S.; He, Z.; Yang, X. Organic soil additives for the remediation of cadmium contaminated soils and their impact on the soil-plant system: A review. *Sci. Total. Environ.* **2020**, *707*, 136121. [CrossRef] [PubMed]
4. Huang, X.; Li, X.; Zheng, L.; Zhang, Y.; Sun, L.; Feng, Y.; Du, J.; Lu, X.; Wang, G. Comprehensive assessment of health and ecological risk of cadmium in agricultural soils across China: A tiered framework. *J. Hazard. Mater.* **2024**, *465*, 133111. [CrossRef] [PubMed]
5. Fuge, R. Fluorine in the environment, a review of its sources and geochemistry. *Appl. Geochem.* **2019**, *100*, 393–406. [CrossRef]
6. Yang, J.-Y.; Wang, M.; Lu, J.; Yang, K.; Wang, K.-P.; Liu, M.; Luo, H.-Q.; Pang, L.-N.; Wang, B. Fluorine in the environment in an endemic fluorosis area in Southwest, China. *Environ. Res.* **2020**, *184*, 109300. [CrossRef] [PubMed]
7. Chen, Y.; Wang, S.; Nan, Z.; Ma, J.; Zang, F.; Li, Y.; Zhang, Q. Effect of fluoride and cadmium stress on the uptake and translocation of fluoride and cadmium and other mineral nutrition elements in radish in single element or co-taminated sierozem. *Environ. Exp. Bot.* **2017**, *134*, 54–61. [CrossRef]
8. Li, Y.; Wang, S.; Prete, D.; Xue, S.; Nan, Z.; Zang, F.; Zhang, Q. Accumulation and interaction of fluoride and cadmium in the soil-wheat plant system from the wastewater irrigated soil of an oasis region in northwest China. *Sci. Total. Environ.* **2017**, *595*, 344–351. [CrossRef] [PubMed]
9. Tang, J.; Xiao, T.; Wang, S.; Lei, J.; Zhang, M.; Gong, Y.; Li, H.; Ning, Z.; He, L. High cadmium concentrations in areas with endemic fluorosis: A serious hidden toxin? *Chemosphere* **2009**, *76*, 300–305. [CrossRef]
10. Xiong, Y.; Xiao, T.; Liu, Y.; Zhu, J.; Ning, Z.; Xiao, Q. Occurrence and mobility of toxic elements in coals from endemic fluorosis areas in the Three Gorges Region, SW China. *Ecotoxicol. Environ. Saf.* **2017**, *144*, 1–10. [CrossRef]
11. McDowell, R.W.; Gray, C.W. Do soil cadmium concentrations decline after phosphate fertiliser application is stopped: A comparison of long-term pasture trials in New Zealand? *Sci. Total. Environ.* **2022**, *804*, 150047. [CrossRef] [PubMed]
12. Haddy, A.; Lee, I.; Shin, K.; Tai, H. Characterization of fluoride inhibition in photosystem II lacking extrinsic PsbP and PsbQ subunits. *J. Photochem. Photobiol. B Biol.* **2018**, *185*, 1–9. [CrossRef] [PubMed]
13. Lu, M.; Yu, S.; Lian, J.; Wang, Q.; He, Z.; Feng, Y.; Yang, X. Physiological and metabolomics responses of two wheat (*Triticum aestivum* L.) genotypes differing in grain cadmium accumulation. *Sci. Total. Environ.* **2021**, *769*, 145345. [CrossRef] [PubMed]
14. Li, J.; Song, Y.; Vogt, R.D.; Liu, Y.; Luo, J.; Li, T. Bioavailability and cytotoxicity of Cerium- (IV), Copper- (II), and Zinc oxide nanoparticles to human intestinal and liver cells through food. *Sci. Total. Environ.* **2020**, *702*, 134700. [CrossRef]
15. Li, Y.; Wang, S.; Zhang, Q.; Zang, F.; Nan, Z.; Sun, H.; Huang, W.; Bao, L. Accumulation, interaction and fractionation of fluoride and cadmium in sierozem and oilseed rape (*Brassica napus* L.) in northwest China. *Plant Physiol. Biochem.* **2018**, *127*, 457–468. [CrossRef] [PubMed]

16. Luo, Y.; Yuan, H.; Zhao, J.; Qi, Y.; Cao, W.-W.; Liu, J.-M.; Guo, W.; Bao, Z.-H. Multiple factors influence bacterial community diversity and composition in soils with rare earth element and heavy metal co-contamination. *Ecotoxicol. Environ. Saf.* **2021**, *225*, 112749. [CrossRef] [PubMed]
17. Tang, C.-C.; Hu, Y.-R.; Zhang, M.; Chen, S.-L.; He, Z.-W.; Li, Z.-H.; Tian, Y.; Wang, X.C.; Tang, C.-C.; Hu, Y.-R.; et al. Role of phosphate in microalgal-bacterial symbiosis system treating wastewater containing heavy metals. *Environ. Pollut.* **2024**, *349*, 123951. [CrossRef] [PubMed]
18. Cui, S.-F.; Yang, J.-Y. Effects of Fluorine on the Growth of Broad Bean (*Vicia faba* L.) and Maize (*Zea mays* L.) and the Response of Microbial Community in Soils. *Water Air Soil Pollut.* **2021**, *232*, 1–14. [CrossRef]
19. Gan, C.D.; Jia, Y.B.; Yang, J.Y. Remediation of fluoride contaminated soil with nano-hydroxyapatite amendment: Response of soil fluoride bioavailability and microbial communities. *J. Hazard. Mater.* **2021**, *405*, 124694. [CrossRef]
20. Lu, M.; Xu, K.; Chen, J. Effect of pyrene and cadmium on microbial activity and community structure in soil. *Chemosphere* **2013**, *91*, 491–497. [CrossRef] [PubMed]
21. Wang, C.; Luo, Y.; Tan, H.; Liu, H.; Xu, F.; Xu, H. Responsiveness change of biochemistry and micro-ecology in alkaline soil under PAHs contamination with or without heavy metal interaction. *Environ. Pollut.* **2020**, *266*, 115296. [CrossRef] [PubMed]
22. Li, J.; Qiu, Y.; Zhao, Q.; Chen, D.; Wu, Z.; Peng, A.-A.; Niazi, N.K.; Trakal, L.; Sakrabani, R.; Gao, B.; et al. Lead and copper-induced hormetic effect and toxicity mechanisms in lettuce (*Lactuca sativa* L.) grown in a contaminated soil. *Sci. Total. Environ.* **2020**, *741*, 140440. [CrossRef] [PubMed]
23. Zare, A.A.; Khoshgoftarmanesh, A.H.; Malakouti, M.J.; Bahrami, H.A.; Chaney, R.L. Root uptake and shoot accumulation of cadmium by lettuce at various Cd:Zn ratios in nutrient solution. *Ecotoxicol. Environ. Saf.* **2018**, *148*, 441–446. [CrossRef] [PubMed]
24. Wu, Y.; Ma, L.; Liu, Q.; Sikder, M.; Vestergård, M.; Zhou, K.; Wang, Q.; Yang, X.; Feng, Y. Pseudomonas fluorescens promote photosynthesis, carbon fixation and cadmium phytoremediation of hyperaccumulator Sedum alfredii. *Sci. Total. Environ.* **2020**, *726*, 138554. [CrossRef]
25. Edwards, J.; Johnson, C.; Santos-Medellín, C.; Lurie, E.; Podishetty, N.K.; Bhatnagar, S.; Eisen, J.A.; Sundaresan, V. Structure, variation, and assembly of the root-associated microbiomes of rice. *Proc. Natl. Acad. Sci. USA* **2015**, *112*, 911–920. [CrossRef] [PubMed]
26. Wang, M.; Zhang, L.; Liu, Y.; Chen, D.; Liu, L.; Li, C.; Kang, K.J.; Wang, L.; He, Z.; Yang, X. Spatial variation and fractionation of fluoride in tobacco-planted soils and leaf fluoride concentration in tobacco in Bijie City, Southwest China. *Environ. Sci. Pollut. Res.* **2021**, *28*, 26112–26123. [CrossRef] [PubMed]
27. Cao, X.; Luo, J.; Wang, X.; Chen, Z.; Liu, G.; Khan, M.B.; Kang, K.J.; Feng, Y.; He, Z.; Yang, X. Responses of soil bacterial community and Cd phytoextraction to a Sedum alfredii-oilseed rape (*Brassica napus* L. and *Brassica juncea* L.) intercropping system. *Sci. Total. Environ.* **2020**, *723*, 138152. [CrossRef] [PubMed]
28. Bolger, A.M.; Lohse, M.; Usadel, B. Trimmomatic: A flexible trimmer for Illumina sequence data. *Bioinformatics* **2014**, *30*, 2114–2120. [CrossRef] [PubMed]
29. Martin, M. Cutadapt removes adapter sequences from high-throughput sequencing reads. *EMBnet J.* **2011**, *17*, 10–12. [CrossRef]
30. Edgar, R.C. UPARSE: Highly accurate OTU sequences from microbial amplicon reads. *Nat. Methods* **2013**, *10*, 996–998. [CrossRef]
31. Edgar, R.C.; Haas, B.J.; Clemente, J.C.; Quince, C.; Knight, R. UCHIME improves sensitivity and speed of chimera detection. *Bioinformatics* **2011**, *27*, 2194–2200. [CrossRef]
32. Bokulich, N.A.; Subramanian, S.; Faith, J.J.; Gevers, D.; Gordon, J.I.; Knight, R.; Mills, D.A.; Caporaso, J.G. Quality-filtering vastly improves diversity estimates from Illumina amplicon sequencing. *Nat. Methods* **2013**, *10*, 57–59. [CrossRef]
33. Caporaso, J.G.; Kuczynski, J.; Stombaugh, J.; Bittinger, K.; Bushman, F.D.; Costello, E.K.; Fierer, N.; Gonzalez Peña, A.; Goodrich, J.K.; Gordon, J.I.; et al. QIIME allows analysis of high-throughput community sequencing data. *Nat. Methods* **2010**, *7*, 335–336. [CrossRef]
34. Dixon, P. VEGAN, a package of R functions for community ecology. *J. Veg. Sci.* **2003**, *14*, 927–930. [CrossRef]
35. Segata, N.; Izard, J.; Waldron, L.; Gevers, D.; Miropolsky, L.; Garrett, W.S.; Huttenhower, C. Metagenomic biomarker discovery and explanation. *Genome Biol.* **2011**, *12*, R60. [CrossRef]
36. Douglas, G.M.; Maffei, V.J.; Zaneveld, J.; Yurgel, S.N.; Brown, J.R.; Taylor, C.M.; Huttenhower, C.; Langille, M.G. PICRUSt2: An improved and extensible approach for metagenome inference. *BioRxiv* **2019**, 672295. [CrossRef]
37. El Rasafi, T.; Oukarroum, A.; Haddioui, A.; Song, H.; Kwon, E.E.; Bolan, N.; Tack, F.M.G.; Sebastian, A.; Prasad, M.N.V.; Rinklebe, J. Cadmium stress in plants: A critical review of the effects, mechanisms, and tolerance strategies. *Crit. Rev. Environ. Sci. Technol.* **2022**, *52*, 675–726. [CrossRef]
38. Khan, M.I.R.; Ashfaque, F.; Chhillar, H.; Irfan, M.; Khan, N.A. The intricacy of silicon, plant growth regulators and other signaling molecules for abiotic stress tolerance: An entrancing crosstalk between stress alleviators. *Plant Physiol. Biochem.* **2021**, *162*, 36–47. [CrossRef] [PubMed]
39. Rizwan, M.; Ali, S.; Adrees, M.; Ibrahim, M.; Tsang, D.C.; Zia-Ur-Rehman, M.; Zahir, Z.A.; Rinklebe, J.; Tack, F.M.; Ok, Y.S. A critical review on effects, tolerance mechanisms and management of cadmium in vegetables. *Chemosphere* **2017**, *182*, 90–105. [CrossRef]
40. Calabrese, E.J.; Agathokleous, E. Accumulator plants and hormesis. *Environ. Pollut.* **2021**, *274*, 116526. [CrossRef]

41. Zorrig, W.; El Khouni, A.; Ghnaya, T.; Davidian, J.-C.; Abdelly, C.; Berthomieu, P. Lettuce (*Lactuca sativa*): A species with a high capacity for cadmium (Cd) accumulation and growth stimulation in the presence of low Cd concentrations. *J. Hortic. Sci. Biotechnol.* **2015**, *88*, 783–789. [CrossRef]
42. Tausta, S.L.; Berbasova, T.; Peverelli, M.; A Strobel, S. The fluoride transporter FLUORIDE EXPORTER (FEX) is the major mechanism of tolerance to fluoride toxicity in plants. *Plant Physiol.* **2021**, *186*, 1143–1158. [CrossRef] [PubMed]
43. Yazdi, M.; Kolahi, M.; Kazemi, E.M.; Barnaby, A.G. Study of the contamination rate and change in growth features of lettuce (*Lactuca sativa* Linn.) in response to cadmium and a survey of its phytochelatin synthase gene. *Ecotoxicol. Environ. Saf.* **2019**, *180*, 295–308. [CrossRef] [PubMed]
44. Yadu, B.; Chandrakar, V.; Keshavkant, S. Responses of plants to fluoride: An overview of oxidative stress and defense mechanisms. *Fluoride* **2016**, *49*, 293–302.
45. Banerjee, A.; Roychoudhury, A. Fluorine: A biohazardous agent for plants and phytoremediation strategies for its removal from the environment. *Biol. Plant.* **2019**, *63*, 104–112. [CrossRef]
46. Kamaluddin, M.; Zwiazek, J.J. Fluoride inhibits root water transport and affects leaf expansion and gas exchange in aspen (*Populus tremuloides*) seedlings. *Physiol. Plant.* **2003**, *117*, 368–375. [CrossRef] [PubMed]
47. Choppala, G.; Saifullah; Bolan, N.; Bibi, S.; Iqbal, M.; Rengel, Z.; Kunhikrishnan, A.; Ashwath, N.; Ok, Y.S. Cellular Mechanisms in Higher Plants Governing Tolerance to Cadmium Toxicity. *Crit. Rev. Plant Sci.* **2014**, *33*, 374–391. [CrossRef]
48. Olszowski, T.; Sikora, M.; Chlubek, D. Combined toxicity of fluoride and cadmium. *Fluoride* **2016**, *49*, 194–203.
49. Li, X.; Meng, D.; Li, J.; Yin, H.; Liu, H.; Liu, X.; Cheng, C.; Xiao, Y.; Liu, Z.; Yan, M. Response of soil microbial communities and microbial interactions to long-term heavy metal contamination. *Environ. Pollut.* **2017**, *231*, 908–917. [CrossRef]
50. Wu, B.; Hou, S.; Peng, D.; Wang, Y.; Wang, C.; Xu, F.; Xu, H. Response of soil micro-ecology to different levels of cadmium in alkaline soil. *Ecotoxicol. Environ. Saf.* **2018**, *166*, 116–122. [CrossRef]
51. Jokanović, S.; Kajan, K.; Perović, S.; Ivanić, M.; Mačić, V.; Orlić, S. Anthropogenic influence on the environmental health along Montenegro coast based on the bacterial and chemical characterization. *Environ. Pollut.* **2021**, *271*, 116383. [CrossRef] [PubMed]
52. Chen, L.; Hao, Z.; Li, K.; Sha, Y.; Wang, E.; Sui, X.; Mi, G.; Tian, C.; Chen, W. Effects of growth-promoting rhizobacteria on maize growth and rhizosphere microbial community under conservation tillage in Northeast China. *Microb. Biotechnol.* **2021**, *14*, 535–550. [CrossRef] [PubMed]
53. Da Silva, W.L.; Yang, K.-T.; Pettis, G.S.; Soares, N.R.; Giorno, R.; Clark, C.A. Flooding-associated soft rot of sweetpotato storage roots caused by distinct Clostridium isolates. *Plant Dis.* **2019**, *103*, 3050–3056. [CrossRef] [PubMed]
54. Araujo, R.; Gupta, V.V.; Reith, F.; Bissett, A.; Mele, P.; Franco, C.M. Biogeography and emerging significance of *Actinobacteria* in Australia and Northern Antarctica soils. *Soil Biol. Biochem.* **2020**, *146*, 107805. [CrossRef]
55. Mitter, E.K.; De Freitas, J.R.; Germida, J.J. Bacterial root microbiome of plants growing in oil sands reclamation covers. *Front. Microbiol.* **2017**, *8*, 849. [CrossRef]
56. Chen, Z.J.; Tian, W.; Li, Y.J.; Sun, L.N.; Chen, Y.; Zhang, H.; Li, Y.Y.; Han, H. Responses of rhizosphere bacterial communities, their functions and their network interactions to Cd stress under phytostabilization by *Miscanthus* spp. *Environ. Pollut.* **2021**, *287*, 117663. [CrossRef]

toxics

MDPI

Article

Quantitative Soil Characterization for Biochar–Cd Adsorption: Machine Learning Prediction Models for Cd Transformation and Immobilization

Muhammad Saqib Rashid [1], Yanhong Wang [1], Yilong Yin [1], Balal Yousaf [2], Shaojun Jiang [1], Adeel Feroz Mirza [3], Bing Chen [4], Xiang Li [1,*] and Zhongzhen Liu [1,*]

1 Key Laboratory of Plant Nutrition and Fertilizer in South Region, Ministry of Agriculture, Guangdong Key Laboratory of Nutrient Cycling and Farmland Conservation, Institute of Agricultural Resources and Environment, Guangdong Academy of Agricultural Sciences, Guangzhou 510640, China; saqibssr@mail.ustc.edu.cn (M.S.R.); wangyanhong@gdaas.cn (Y.W.); yilong516@163.com (Y.Y.); shaojunj93@163.com (S.J.)

2 Department of Technologies and Installations for Waste Management, Faculty of Energy and Environmental Engineering, Silesian University of Technology, 44-100 Gliwice, Poland; balal.yousaf@polsl.pl

3 Department of Mechanical and Energy Engineering, Southern University of Science and Technology, Shenzhen 518055, China; adeelmirza@mail.ustc.edu.cn

4 Key Laboratory of Animal Nutrition and Feed Science in South China, Guangdong Provincial Key Laboratory of Animal Breeding and Nutrition, Collaborative Innovation Center of Aquatic Sciences, Institute of Animal Science, Guangdong Academy of Agricultural Sciences, Ministry of Agriculture and Rural Affairs, Guangzhou 510640, China; chenbing114@163.com

* Correspondence: lzzgz2001@163.com (Z.L.); lixiang142213@163.com (X.L.)

Abstract: Soil pollution with cadmium (Cd) poses serious health and environmental consequences. The study investigated the incubation of several soil samples and conducted quantitative soil characterization to assess the influence of biochar (BC) on Cd adsorption. The aim was to develop predictive models for Cd concentrations using statistical and modeling approaches dependent on soil characteristics. The potential risk linked to the transformation and immobilization of Cd adsorption by BC in the soil could be conservatively assessed by pH, clay, cation exchange capacity, organic carbon, and electrical conductivity. In this study, Long Short-Term Memory (LSTM), Bidirectional Gated Recurrent Unit (BiGRU), and 5-layer CNN Convolutional Neural Networks (CNNs) were applied for risk assessments to establish a framework for evaluating Cd risk in BC amended soils to predict Cd transformation. In the case of control soils (CK), the BiGRU model showed commendable performance, with an R^2 value of 0.85, indicating an approximate 85.37% variance in the actual Cd. The LSTM model, which incorporates sequence data, produced less accurate results ($R^2 = 0.84$), while the 5-layer CNN model had an R^2 value of 0.91, indicating that the CNN model could account for over 91% of the variation in actual Cd levels. In the case of BC-applied soils, the BiGRU model demonstrated a strong correlation between predicted and actual values with R^2 (0.93), indicating that the model explained 93.21% of the variance in Cd concentrations. Similarly, the LSTM model showed a notable increase in performance with BC-treated soil data. The R^2 value for this model stands at a robust R^2 (0.94), reflecting its enhanced ability to predict Cd levels with BC incorporation. Outperforming both recurrent models, the 5-layer CNN model attained the highest precision with an R^2 value of 0.95, suggesting that 95.58% of the variance in the actual Cd data can be explained by the CNN model's predictions in BC-amended soils. Consequently, this study suggests developing ecological soil remediation strategies that can effectively manage heavy metal pollution in soils for environmental sustainability.

Keywords: biochar; cadmium; transformation; remediation; prediction models

1. Introduction

Natural and synthetic processes release heavy metals (HMs) into the environment, including volcanic eruptions, weathering, wastewater irrigation, sewage sludge disposal, smelting, and pesticide application [1,2]. Furthermore, ingestion and inhalation absorption are the main routes by which HMs can accumulate in the human body [3]. The soil–crop system provides an additional route by which HMs can accumulate in humans [4–6]. In organisms, Cd can accumulate for 50 years, and its half-life is 10 to 30 years. According to a 2019 Environmental Protection Agency (EPA) report, Cd and its compounds were considered hazardous metals and toxic water pollutants [7]. Approximately 7.0% of the soils of China was found to contain excess Cd, which ranked first among inorganic pollutants [8].

The physiochemical composition and texture of soil determine its quality and production [9]. These characteristics form the foundation of the natural environment and crops [10]. HM pollution is a significant by-product of industrialization in various countries worldwide [11]. China's rapid industrialization has inevitably produced similar problems, and governments at all levels are committed to remediating HM-contaminated soils [12]. In addition to microbial activity, organic matter can contribute to the immobilization of Cd through precipitation and complexation processes in soil [13,14]. Standard methods used for HM analysis include ultraviolet-visible spectrophotometry, gas chromatography–mass spectrometry, inductively coupled plasma emission spectrometry, and atomic fluorescence spectrometry [15]. While these technologies possess high sensitivity and accuracy, their short detection range, labor-intensive and time-consuming detection techniques, and complex preparation of samples contribute to their being unsuitable for quick, non-intrusive, and batching testing [16,17].

In recent years, artificial intelligence (AI) deep learning approaches have conquered the limitations of conventional diagnosing methods [18]. Such approaches improve efficiency and adaptability by reducing the amount of initial processing and incorporating the required extract [19]. Deep learning is a hierarchical structure that uses machine learning techniques to identify and collect significant characteristics [20]. It improves its performance via training, resulting in high precision. Statistical and CNN approaches estimate tangible qualities from the input [21]. Among the emerging soil amendment materials in recent years, BC has been regarded as one of the most effective [22]. Several studies have demonstrated that BC, a charcoal derived from organic matter, is highly effective as an adsorbent for Cd in soil [23]. It has been found that clay minerals and organic matter in soil can enhance BC's performance in the adsorption of Cd [24]. Consequently, clay minerals and organic matter increase the soil's surface area. This leads to an increase in the surface area accessible for the adsorption of Cd and facilitates the elimination of Cd from the soil [25]. The highly porous structure of BC increases the surface area [26], which could contribute to the adsorption of Cd through increased adsorption sites and surface complexation [27]. Due to ion exchange sites in clay minerals and organic matter, Cd cannot leach into groundwater or be absorbed by plants [28].

BC can impact the adsorption of Cd through the processes of chemical bonding, pH modification, and microbial activity, resulting in the formation of less toxic forms [29]. Learning complicated representations at different levels of abstraction is made possible by layer-wise layering of several nonlinear and linear computational modules [30]. Researchers have developed predictive algorithms that minimize errors in prediction versus experimental data for evaluating substance qualities [31]. Learning algorithms are developed using marked or unmarked data, or mixtures of data, to generate artificial intelligence algorithms. Data scientists utilize several machine-learning techniques for categorization and predictive modeling [32,33]. However, using quantitative analysis techniques to characterize soil's properties and BC's influence on Cd adsorption using advanced prediction methods is still unclear. Given the above problems, there is still a lack of quantitative and general guiding research, which limits the precise application of BC in improving HM-contaminated soil using advanced prediction models.

To assess soil sensitivity, the research assembled representative soil samples across the country with an extensive range of physicochemical properties. The present study assessed the potential of Cd bioaccumulation in soils to evaluate the efficacy of several prediction models, including LSTM, BiGRU, and 5-layer CNN, for risk assessments. The objective was to provide an analytical approach to evaluating the impact of BC addition on Cd retention in soil characteristics, including pH, OC, CEC, EC, clay, and P. Additionally, AI algorithms were used to estimate the initial basic associations of soil parameters to enhance prediction models for BC-Cd soils and explore the contributing factors of Cd immobilization in soils. The analytical methodologies used in this investigation consisted of internationally recognized standards and innovative technologies that have emerged in recent years.

2. Materials and Methods

2.1. Soil Samples Description

This study was intended to select 44 soil samples from different areas of China (Table S1). The soil samples were collected from the upper layer of about 20 cm of the soil profile and brought to the Guangdong Academy of Agricultural Sciences. A composite sample was formed by thoroughly blending and homogenizing each subsample. After air-drying, a 2 mm sieve was used to collect debris from the composite soil sample [34].

2.2. Biochar Preparation

Rice straw biomass was purchased from the local commercial market and washed with distilled water (10–15 s) to remove dust particles. Subsequently, biomass was air-dried at room temperature and then oven-dried overnight with a constant hot air supply at 105 °C to remove moisture, after which it was mechanically ground. Rice straw BC at 450 °C was prepared with a retention time of 2 h. Afterward, it was cooled and passed through a 1 mm sieve [35]. The basic properties of BC are given in Table S2.

2.3. Incubation Experiment

An incubation experiment was conducted to evaluate the physicochemical characteristics of the soil by adding BC to enhance the adsorption of Cd. The significance of soil parameters and the sensitivity of BC-Cd adsorption were examined in various types of soils. Analytical-grade chemicals were employed. All chemical reagents were purchased from Sigma-Aldrich (Shanghai, China) and Sinopharm Chemical Reagent Company Ltd., Shanghai, China. De-ionized water (ultrapure) was used. Cd-contaminated soils were prepared artificially with $Cd(NO_3)_2$ (purity of 99.9%) aqueous solution (1000 mg/L). The purpose of soil spiking was to achieve a target concentration of about 3.3 mg/kg. The spiked soils were kept for 15 days to equilibrate Cd-contaminated soils. Afterwards, BC amendment was applied to soil at a rate of 1% (*w*/*w*). The moisture was maintained at 60–70%. The total duration of this experiment was 45 days. Before and after the incubation experiment's completion, soil physicochemical analysis, which evaluated the pH (1:2.5 (*w*/*v*)) using Mettler Toledo (Chengdu, China, model: Seven Compact 8210); the OC using the volumetric method (titration and colorimetric) by [36]; the EC (1:2.5 (*w*/*v*)) using FOSS (model: TFS/YS-203) from Shanghai Hongyi Instrumentation Co., Ltd. (Shanghai, China); CEC using the ammonium acetate extracts method; and the total Cd (mg/kg) using Agilent Technologies (Chengdu, China) model: 7800 ICP-MS (inductively coupled plasma mass spectrometry), as well as tri acids (HNO_3:H_2SO_4:$HClO_4$) [2]. For soil texture, 40 mL of 1% sodium hexametaphosphate was applied to the soil (40 g). After waiting overnight, the soil was shifted to a dispersion cup (mechanical stirrer). The reading was taken using a Bouyoucos hydrometer. The texture of the soil was evaluated using the textural triangle. For diethylenetriaminepentaacetic acid (DPTA) analysis, the soil (8 g) was weighed, and an extraction solution of 16 mL of DTPA (pH 7.3) was added. After 60 min of shaking, the sample was filtered, and Cd was measured using ICP-MS [37].

2.4. Proposed Machine Learning Methods

2.4.1. LSTM Model

LSTM models are specialized recurrent neural networks (RNN) capable of learning long-term dependencies in sequence data (Figure 1). Unlike standard feedforward neural networks, LSTMs have feedback connections that make them powerful for processing single data points and entire data sequences. A key feature of LSTM units is their ability to remember information for long periods, which is achieved through a complex mechanism of gates, including input, forget, and output gates. These gates effectively allow the network to add information to or remove it from the cell state, which is carefully regulated to prevent the vanishing gradient problem often encountered in traditional RNNs. LSTMs are widely used in various applications, such as time series prediction, natural language processing, and sequence generation tasks [38].

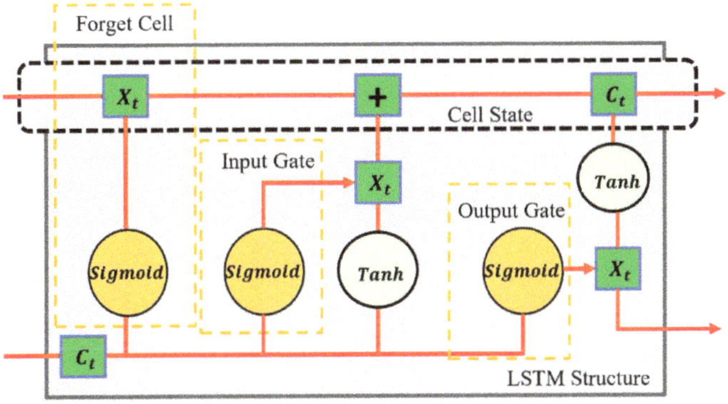

Figure 1. Long Short-Term Memory (LSTM).

The LSTM model was applied to estimate the concentration of Cd (mg/kg) with various soil properties as input parameters. These properties included sand%, clay%, silt%, EC measured in µS/cm, pH, OC content in g/kg, CEC in cmol+/kg, and available P in mg/kg. The choice of LSTM for this task leverages its ability to process and learn from the sequential or structured nature of the input data, even though soil parameters are not sequential in the traditional sense. The model evaluated complex relationships and interactions between these soil properties to predict the Cd levels accurately. The sequential examination of various soil properties using LSTM has the potential to reveal patterns of Cd availability that are essential to environmental monitoring and agriculture. The LSTM's architecture, including its memory cells and gates, can effectively learn from the intricacies of soil data, providing a powerful tool for predicting the HM contamination in soils.

2.4.2. BiGRU Model

In this research, BiGRU networks were employed to estimate Cd concentration based on a set of soil parameters. Figure 2 presents the block diagram of BiGRU. BiGRU models, an advancement of traditional GRU networks, process data in both forward and backward directions, allowing them to capture dependencies and patterns that might be missed when data are processed in a single direction. This dual-direction processing is particularly beneficial for understanding the complex interactions between soil properties and their impact on Cd availability. The study's utilization of BiGRU networks enhanced the efficiency of parameterization and increased the ability to simulate the spatial and temporal relationships between BC-amended soils and Cd. Unlike LSTMs, GRUs simplify the gating mechanism without compromising the model's ability to manage long-term dependencies. The bidirectional nature of BiGRUs offers a comprehensive perspective on the data, ensuring that the temporal dynamics and interdependencies of soil characteristics are thoroughly analyzed. This methodological approach underscores the potential of BiGRU models in environmental sciences, particularly for predicting HM concentrations in soils, thus providing valuable insights for soil management and remediation strategies. The use of BiGRU models highlights the innovative application of deep learning techniques in tackling complex environmental challenges [39].

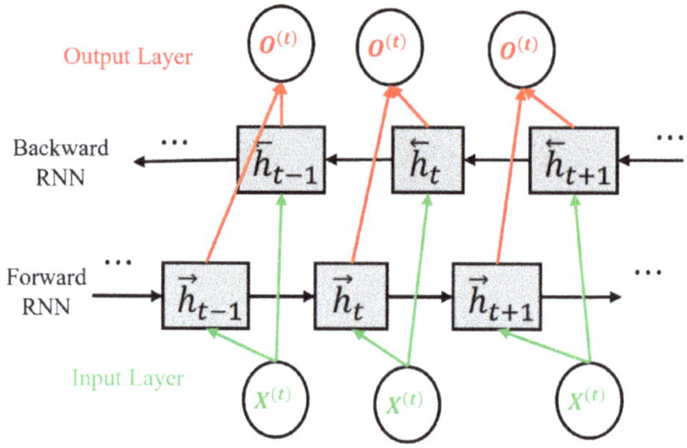

Figure 2. Bidirectional Gated Recurrent Unit (BiGRU).

2.4.3. 5-Layer CNN

This study also explored CNN use by integrating a 5-layer CNN model equipped with max-pooling and fully connected (FC) layers. CNNs, known for image processing and machine learning, can computationally and adaptively learn spatial feature frameworks from the input for the purpose of analyzing complex, multivariate environmental data. The structure of the employed CNN model comprises five convolutional layers, each followed by a max pooling layer. Convolutional layers act as feature extractors from the input data, using learnable filters to identify and capture patterns such as edges, textures, or more complex features in deeper layers. After each convolutional operation, max-pooling layers are applied to reduce the dimensionality of the feature maps. This operation helps to make the representation smaller and more manageable and introduces translational invariance to the features, making the model more robust. The sequence of the convolutional and max pooling layers is designed to progressively refine the feature maps, ensuring that only the most relevant spatial features are retained and highlighted. A fully connected (FC) layer incorporates high-level, filtered data into the prediction model after decoding and merging layers are applied. The FC layer serves as a classifier, mapping the learned features to the output, which in this case is the estimated concentration of Cd [40].

Figure 3 shows the structure of the 5-layer CNN model. Applying a 5-layer CNN model to estimate Cd concentrations from soil parameters is innovative, as it transfers profound learning principles from their conventional domains to environmental science. By adapting CNN architectures, known for their efficiency in handling spatial data, to analyze and learn from soil property data, this study opens new pathways for advanced soil contamination analysis. The CNN model's ability to discern intricate patterns within complex datasets could provide more accurate and reliable predictions of HM contamination, offering significant contributions to environmental monitoring and soil health assessment strategies.

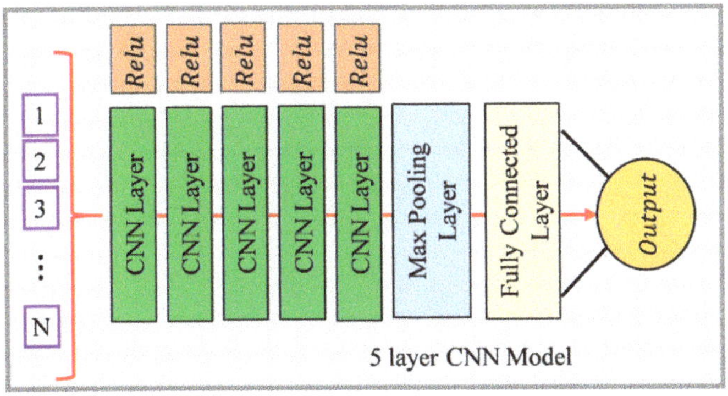

Figure 3. 5-layer CNN model (Convolutional Neural Network).

2.5. Statistical Analysis

All data, including the soil's physiochemical and amendment parameters, were derived from replications. A significance level of $p < 0.05$ was established for the descriptive data using a 95% confidence interval. The tables and graphical representation were generated using Sigma Plot 15.0, Microsoft Excel 2021, and Origin Pro 8.5. The experimental simulation for this study used AMD CPU Ryzen R7-4800H, along with an NVIDIA GeForce RTX 2060 6 GB GPU operating on Windows 10. The neural network model was developed using Python 3.8 and implemented using the Keras framework, with TensorFlow-GPU serving as the backend for deep learning computations.

3. Results and Discussion

3.1. Effect of the Amendment on Incubation Soil Physiochemical Characteristics Relates to Cd Concentrations

BC can immobilize HMs in soils, potentially aiding in the remediation of contaminated soils [41]. Solubility in the soil is attributed to precipitating, redox reactions, formation of complexes, and adsorption processes, which are facilitated by binding HMs to soil components and minerals [26]. The effect of BC additives on the physiochemical characteristics of incubated soils and Cd availability are detailed below.

3.1.1. Soil pH Affects Cd Availability

The influence of applied BC on soil pH and Cd availability is given in Figure 4a. Soil pH was divided into three categories: (i) pH < 6.25, (ii) pH 6.25–7.75, and (iii) pH > 7.75. According to Figure 4a, the results indicated that the concentrations of available Cd were decreased in applied BC soils in all three categories of pH as compared to the control. The results revealed that applied amendment significantly decreased Cd availability in soils pH > 7.75 compared to pH < 6.25 and pH 6.25–7.75. Meanwhile, Figure 4a depicts that the highest Cd availability among control soils was observed at pH < 6.25. BC can alter several soil physicochemical properties owing to its varied composition and biodiversity activities [42]. BC usually has a pH range of 7–10, indicating an alkaline nature. BC is generated due to the pyrolysis of biomass, which has a higher pH than the original biomass [43]. Decontamination of HMs involves surface functional groups consisting of oxygen and the formation of complexes with BC, including electrostatic interactions, exchange of ions, and precipitation of chemicals [44]. Carbon sources, the formation of minerals, hydroxyl ion generation, and basic cation discharge may all be attributed to an increase in pH due to organic matter [10]. Reducing soil acidity and decelerating soil acidification are essential for preventing Cd accumulation in soil systems. Enhancing the buffering ability of soil pH can efficiently inhibit the process of soil acidification, diminish the concentration of accessible HMs in acidic soils, and hinder the soil–plant system absorption of HMs [45,46].

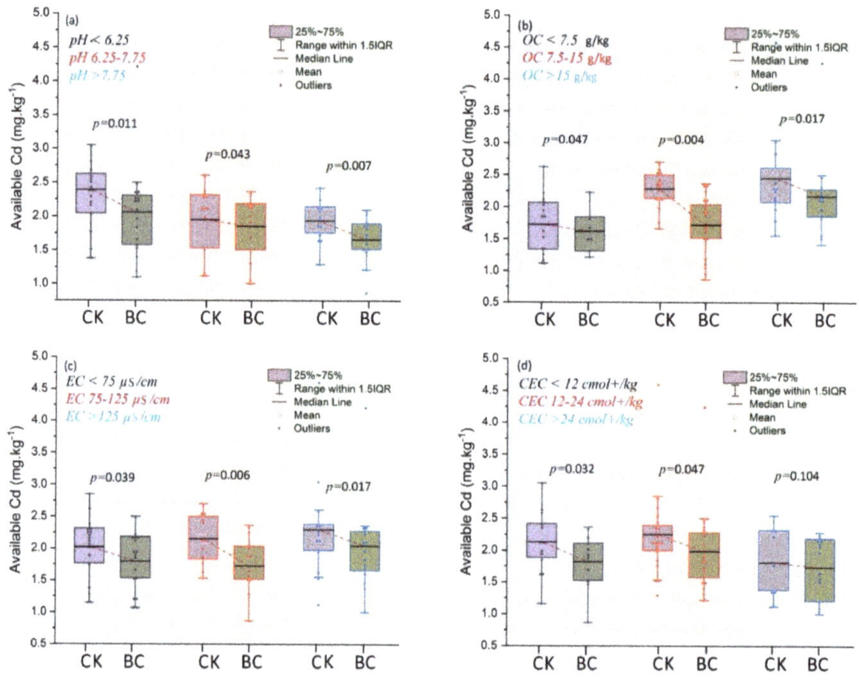

Figure 4. Effect of BC amendment on incubated soil physiochemical properties and Cd (mg/kg) availability. (**a**) pH, (**b**) OC, (**c**) EC, and (**d**) CEC.

3.1.2. Changing Soil Organic Carbon Affects Cd Availability

A comparison of BC amended and un-amended soils on organic carbon (OC) and Cd availability is given in Figure 4b. All incubated soils (amended and un-amended) were divided into three categories based on different ranges, such as low (OC < 7.5 g/kg), medium (OC 7.5–15 g/kg), and high (OC > 15 g/kg). The findings showed that adding BC to all soil samples significantly reduced the availability of Cd compared to the control. According to the findings, the lowest Cd availability was found in the OC (<7.5 g/kg) range of BC-amended soils compared to the control. The result indicated that the inclusion of BC systematically decreases the concentration of Cd in soils at each OC level compared to CK soils. BC minerals or surface groups illustrate inorganic components that mainly contribute to the adsorption of Cd. The findings indicated that the presence of BC can reduce the availability of Cd by binding to it or influencing the characteristics of the soil, thereby mitigating the impact of organic material solely on Cd mobility in the soil. The results demonstrated that using BC may enhance soil carbon sequestration and nutrient retention. It was shown that BC substantially increased the organic matter content, which was indirectly linked to changes in residual Cd concentration [47]. Organic matter is a significant component of soil fertility, providing nutrients for sustainable agriculture to maintain soil fertility. The OC level in the soil is essential for retaining nutrients and water, strengthening the soil structure, and supplying energy to soil microorganisms [48]. BC enhances soil fertility through the following mechanisms: retention of fertilizers, stimulation of microbial activity, carbon dioxide emission reduction, and immobilization of organic and inorganic pollutants [49]. The extensive integration of BC into soils would inevitably impact the increase in the quantity and composition of soil OC [50]. The effective sorption of HMs in contaminated effluents and soils has been associated with BC, which can potentially eradicate environmental contamination on a global scale [51].

BC possesses a substantial quantity of organic carbon, making it an exceptional asset in soil particle aggregation. Soil pollutants were immobilized by modifying soil physicochemical qualities, including carbon sequestration and fertility, with BC applied to the polluted soil [52]. The high area-to-volume ratio of these plants contributes to soil structure, porosity, water-holding capacity, aeration, nutrient availability, and buffering soil pH [53]. Moreover, they can retain plant nutrients, such as calcium, potassium, and phosphorus, reducing leaching [54]. Organic matter increases soil structure, porosity, and water-holding capacity, contributing significantly to climate change mitigation and carbon sequestration [55,56]. By providing nutrients to the soil, it promotes nutrient cycling through the action of soil microorganisms and serves as a source of nutrients. Additionally, it decomposes into stable aggregates that support soil structure, which supports soil porosity and improves soil structure [57].

3.1.3. Soils EC Affects Cd Availability

The comparison of amended and un-amended soil EC (µS/cm) and Cd availability is shown in Figure 4c. EC (µS/cm) was divided into three different ranges: (i) EC < 75 µS/cm; (ii) EC 75–125 µS/cm; and (iii) EC > 125 µS/cm. Soil ecosystems are intricate, and the impact of added BC on soil EC-Cd availability can fluctuate based on various variables, including ecological factors, BC quantity, Cd concentration, and the characteristics of the soil. Figure 4c illustrates a particular set of scenarios in which the availability of Cd is substantially influenced by the increase in BC-added soil EC, but only in comparison to CK soils. The lowest Cd availability was found in EC (75–125 µS/cm) in BC-amended soil compared to the control. However, the results indicated that all unamended soils increased Cd availability with an increase in the EC range. EC is an indication of soil properties and also shows the soil water's electrical carrying capacity. The low EC levels indicate limited nutrient availability and organic matter concentration. This finding further confirms the consistent relationship between the variance in BC and EC contents and the effect of Cd availability. The EC of soils has been shown to have a strong link with the movement of metals inside the soil. Soil salinization and acidity are prevalent and substantially affect Cd availability in soil systems [58,59]. BC played a crucial role in lowering the presence of Cd in the polluted soil. It achieved this by raising the soil's pH, OC, EC, and P availability. Additionally, BC reduced the amount of Cd that can be easily exchanged in the soil by changing it into less accessible forms. As a result, there was a considerable decrease in Cd levels [60,61]. Excessive EC showed excess saline and inadequate nutrients, inhibiting development and appropriate plant growth. This may lead to soil–plant toxicity. As measured by EC, the ideal range for soil–plant nutrients is 0.2–1.2 (dS/m). However, adding BC may enhance soil EC due to the substantial amount of dissolved salts it contains. Recent research has shown that BC can increase the EC in soil [62].

3.1.4. Changing Soil CEC Affects Cd Availability

A comparison of the effects of amended and un-amended soil CEC on Cd availability is presented in Figure 4d. Soil CEC (cmol+/kg) was also divided into 3 different ranges (i) CEC < 12 cmol+/kg; (ii) CEC 12–24 cmol+/kg; and (iii) CEC > 24 cmol+/kg. The results revealed that Cd availability decreased and was observed in all amended soils. The findings indicated that BC amendment significantly reduced Cd availability in soils compared to the control. The results showed that CEC indirectly correlated with Cd availability compared to the control. Application of BC led to enhancements in soil characteristics and resulted in the transformation of the Cd in the soil into more stable portions [63,64]. The rise in CEC may have led to the soil producing significant quantities of quinones, phenols, and carbonyls, thus resulting in greater adsorption of Cd [65,66].

The efficacy of immobilization is affected by several processes and circumstances, including precipitation, electrostatic contact, cation exchange, and complexation with functional groups [67,68]. A high CEC and a high adsorption capacity are both characteristics of BC which may have either negative or positive charges on its interface. The application of BC to enhance soil CEC resulted in a reduction in Cd through ion exchange and more efficient complexation adsorption into the soil [69]. Because soil clay particles assimilated more Cd as the CEC increases, the accessible Cd level in the soil was reduced significantly [70]. BC-amended soil exhibited an increase in CEC, which was dependent on the OC and clay along the surface area and electronegativity for ion adsorption sites capable of immobilizing Cd [71]. CEC was the most critical parameter significantly impacting Cd bioavailability and fractionation in soil [72].

3.1.5. Effect of Soil Clay on Cd Availability

A comparison of amended and un-amended soil clay on Cd availability is shown in Figure 5. Soil clay was divided into three different categories: clay < 20%, clay 20–40%, and clay > 40%. According to Figure 5, the results showed that the BC amendment significantly decreased Cd availability compared to the control. The lowest Cd availability (almost 40–60% reduction) was found in high-clay-content, BC-amended soil compared to the control. However, the results indicated that all unamended soils had increased Cd availability along with the increased clay content. Minerals such as clay are formed from various elements, such as aluminum, silica, iron, etc. [73]. This finding further confirms the synergic relationship between BC and soil clay content compared to the control. The results revealed that soil clay was shown to have a strong link with the movement of Cd availability in the soil. The finding indicated that soil clay's direct relationship with Cd availability and applied BC led to enhancements in soil characteristics and resulted in the immobilization of Cd in the soil. BC also reduced Cd availability by increasing pH, which increased the clay mineral surface negative charges and Cd ion adsorption [24]. Adding BC to medium or coarse-textured soils may improve the consistency of aggregate size and cationic retention.

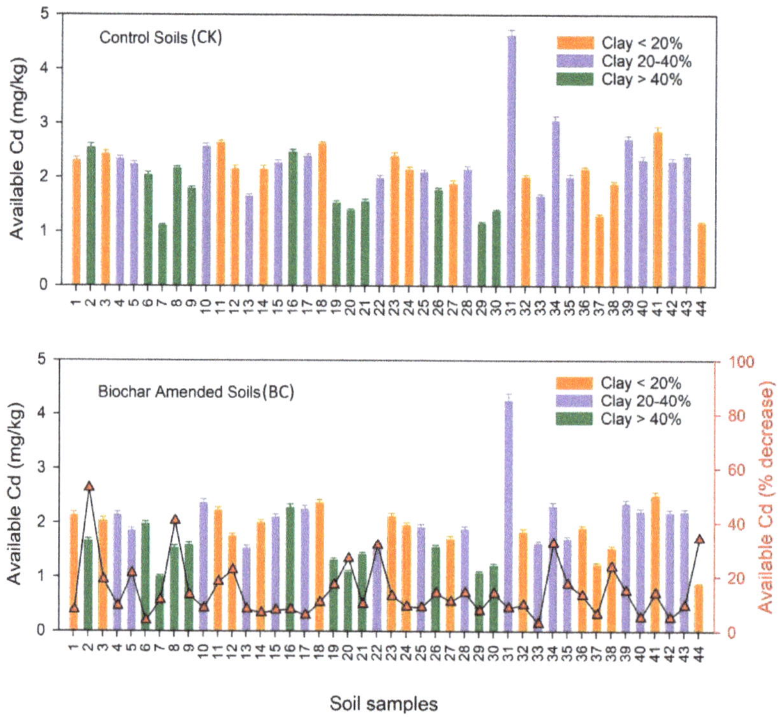

Figure 5. Impact of BC amendment on incubated soil clay and Cd (mg/kg) availability.

Clay minerals and organic matter are critical components of soil that significantly impact the soil's health, plant growth, and environmental sustainability [74,75]. Alumina, silica, and other elements constitute earth crust clays, and these soil metal sorbents reduce metal bioavailability [74]. Clay minerals' exchangeable cations and anions contribute to their efficient contamination mitigation tools. Clays, which are negatively charged, absorb metal from soil solutions. Clay minerals use ion exchange, complexation, and direct bonding to absorb HMs [75]. Natural clay minerals have a strongly negatively charged layer to absorb cations. Minerals' hydroxyl groups adsorb or combine HM to reduce its availability. In soil profiles, this adsorption reduces HM leaching. HM contaminants in soil have been successfully fixed in situ via adsorption and the formation of low-solubility HM precipitates using organic materials, including BC, clay minerals, and functional adsorbents [76]. Its considerable ecological and agronomic advantages render BC a highly prospective soil remediation agent, which has led to its widespread acceptance as a soil amendment.

3.2. Machine Learning-Based Soil Cd Concentration Prediction Models

The key goal of this research was to obtain representative soil samples from different regions of the country, each with distinct physical and chemical characteristics, to investigate the response of soils to the inclusion of BC and establish a prediction framework for the transformation and immobilization of Cd. The comprehensive analysis of the predictive modeling approaches was employed to estimate the concentration of Cd in soil based on an extensive dataset encompassing a wide array of soil properties, including sand%, clay%, silt%, EC, pH, OC, CEC, and available P. The significance of this study is underpinned by the environmental and agricultural importance of accurately assessing Cd levels in soils, given Cd's known adverse effects on plant growth, soil health, and, ultimately, food safety. Cd is a considerable environmental concern due to its persistence and bioaccumulation in ecosystems, making its estimation in agricultural soils crucial for ensuring crop safety and ecological health.

The LSTM model was used to estimate the concentration of Cd from soil properties like sand%, clay%, silt%, EC, pH, OC content, CEC, and available P. The model learned from the sequential nature of input data, evaluating complex relationships between soil properties to predict Cd levels accurately. This sequential examination of soil properties can reveal patterns of Cd availability that are essential for environmental monitoring and agriculture. The LSTM's architecture, including memory cells and gates, can effectively learn from soil data, making it a powerful tool for predicting HM contamination in soils. The research utilized BiGRU networks to estimate the concentration of Cd in soil parameters. BiGRU models process data in both forward and backward directions, capturing dependencies and patterns that might be missed when processed in a single direction. This dual-direction processing helps us to understand complex interactions between soil properties and their impact on Cd availability. BiGRU networks enhance parameterization efficiency and simulate spatial and temporal relationships between BC-amended soils and Cd. This methodological approach highlights the potential of BiGRU models in environmental sciences, particularly for predicting HM concentrations in soils, providing valuable insights for soil management and remediation strategies. The study used a 5-layer CNN model, which included max-pooling layers and a fully connected layer, to analyze complex environmental data. The model consisted of five convolutional layers, each with a max pooling layer, which extracted patterns from the input data. The model's sequence refined the feature maps, ensuring that only relevant spatial features were retained. The fully connected layer incorporated high-level, filtered data into the prediction model, mapping learned features to the output, such as the estimated concentration of Cd in soil parameters. This innovative approach could contribute to environmental monitoring and soil health assessment strategies. The correlation heatmap was generated based on the dataset (Figures 6 and 7). It provides a visual summary of the distributions of the soil properties and illustrates each feature's spread and central tendency. The heatmap shows the correlation coefficients between all pairs of numerical variables, giving insight into the relationships between soil properties and Cd levels. In the case of CK soils, the findings showed positive relationships (Figure 6) between Cd content and silt% (0.42), organic carbon (0.39), EC (0.15), and P (0.20). The correlation matrices of the data of CK soils and Cd showed that there was a negative correlation with clay% (−0.32), sand% (−0.11), pH (−0.30), and CEC (−0.23). However, for soils modified with BC (Figure 7), the findings showed further evidence via negative relationships with sand% (−0.08), clay% (−0.35), pH (−0.25), and CEC (−0.24). The positive correlation matrices of BC-treated soils' cumulative data revealed that the Cd levels were positively linked with the silt% (0.43), EC (0.13), OC (0.42), and P (0.25) [77]. The comprehensive details are given in Section 3.1.

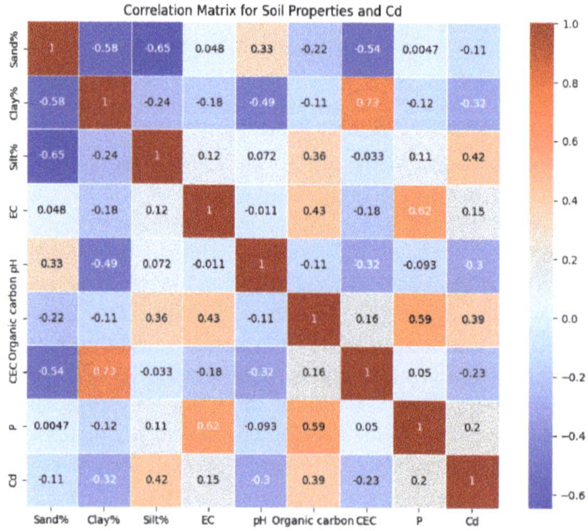

Figure 6. Correlation matrix for soil properties of CK.

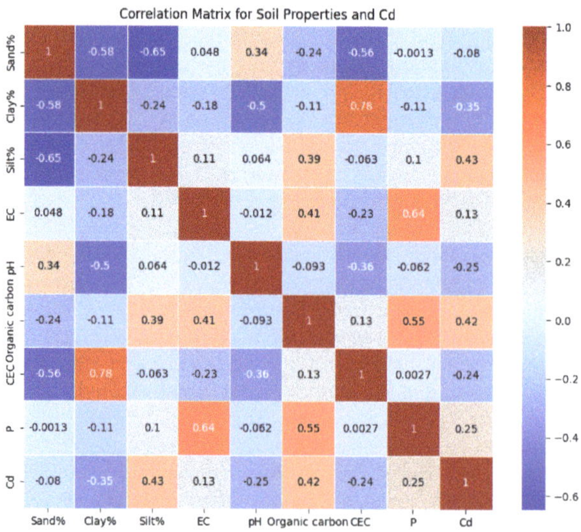

Figure 7. Correlation matrix of soil properties with BC.

3.2.1. Case 1: Assess CK Soils Using Machine Learning Models

Figure 8a compares the actual Cd concentrations against the predictions made by three machine learning models. This comparison is made in the context of soil samples without adding BC. The BiGRU model showed commendable performance, with an R^2 value of 0.8537, indicating that the model's predictions could explain approximately 85.37% of the variance in the actual Cd data. The model's mean absolute error (MAE) was reported at 0.1986, and the mean squared error (MSE) was 0.0650, with a root mean square error (RMSE) of 0.2550. These metrics suggest that the BiGRU model predictions were relatively close to the actual values, with a moderate spread of errors. The LSTM model, which also considered sequence data, yielded slightly less accurate results, with an R^2 value of 0.8473. The MAE for this model was 0.2127, the MSE was 0.0678, and the RMSE was 0.2605, which are all slightly higher than those for the BiGRU model, indicating a lower performance in terms of capturing the variation in Cd concentrations. The 5-layer CNN model, traditionally used in image processing but adapted here for sequence prediction, outperformed both recurrent neural network models with an R^2 value of 0.9107, showing that over 91% of the variability in the actual Cd levels could be accounted for by the CNN model's predictions. Additionally, this model achieved the lowest MAE of 0.1602, the lowest MSE of 0.0439, and the lowest RMSE of 0.2121, indicating the highest accuracy and precision among the three models.

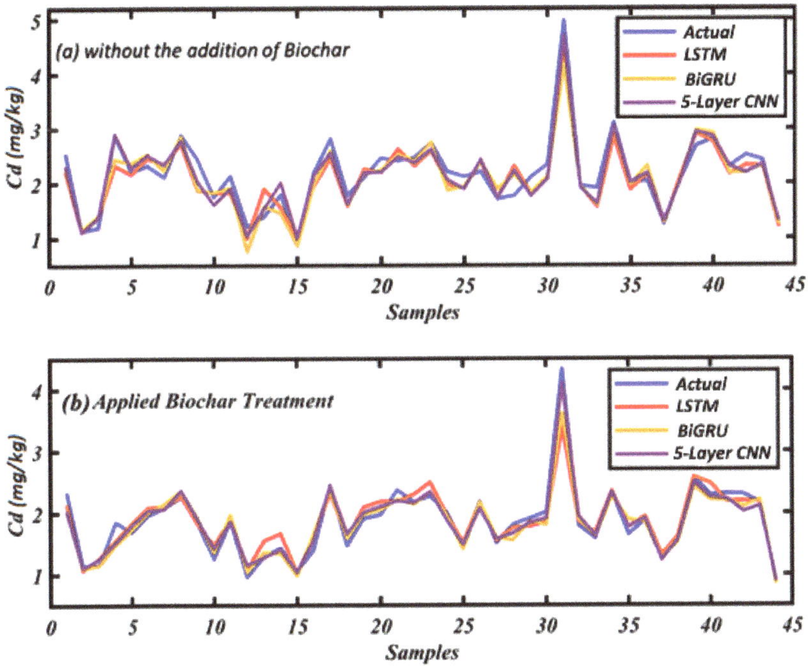

Figure 8. The effectiveness of machine learning models in estimating Cd.

Overall, Figure 8a demonstrates that, while all models were relatively proficient in estimating Cd concentrations, the 5-layer CNN model appeared to be more effective in this context. This might be due to its superior ability to capture the complex non-linear relationships between the soil properties and the Cd levels without the influence of BC. The success of the CNN model here suggests that it may be a promising tool for environmental monitoring applications where understanding the baseline contamination levels is crucial.

3.2.2. Case 2: Evaluated Applied BC Soils Using Machine Learning Models

Figure 8b showcases the effectiveness of machine learning models in estimating Cd concentrations in soil, considering the addition of BC. BC, an amendment known for its ability to absorb and immobilize HMs, provides a unique context for the machine learning models to capture the altered relationships between soil properties and Cd availability. The BiGRU model demonstrated a strong correlation between predicted and actual values, with an R^2 value of 0.9321, indicating that the model explained 93.21% of the variance in Cd concentrations. The model improved its predictive accuracy in the presence of BC, with an MAE of 0.1270, an MSE of 0.0165, and an RMSE of 0.1286, all lower than the values reported in Case 1. Similarly, the LSTM model showed a notable increase in performance with BC-treated soil data, yielding an MAE of 0.1105, an MSE of 0.0161, and an RMSE of 0.1256. The R^2 value for this model stood at a robust 0.9413, reflecting its enhanced ability to predict Cd levels with BC incorporation. Outperforming both recurrent models, the 5-layer CNN model attained the highest precision, with an R^2 value of 0.9558, suggesting that 95.58% of the variance in the actual Cd data could be explained by the CNN model's predictions in BC-amended soils. The model registered the lowest error metrics, with an MAE of 0.0983, an MSE of 0.0145, and an RMSE of 0.1206, underlining its superior predictive power in this scenario.

The results in Figure 8b reinforce the potential of advanced deep learning models in environmental applications, especially for assessing the impacts of soil amendments such as BC on HM contamination. The improved accuracy and lower error rates across all models with the addition of BC suggest that these models can capture the complex interactions that BC induces in the soil matrix, subsequently affecting Cd bioavailability. These insights are valuable for designing strategies to mitigate soil contamination and evaluating BC's effectiveness as a remediation tool. The superior performance of the 5-layer CNN model, in particular, highlights the adaptability and robustness of convolutional networks in handling diverse and complex environmental datasets.

4. Conclusions

The study compared the effectiveness of deep learning machines like LSTM, BiGRU, and 5-layer CNN in forecasting Cd concentrations. A significant negative association between the two variables was found, mainly due to the impact of BC on soil characteristics and Cd immobilization. BC effectively mitigates the presence of HMs in soil and inhibits their penetration into the soil system. The BiGRU model explained 93.21% of the Cd concentration variance after adding BC, while the LSTM model improved with BC-treated soil data. The 5-layer CNN model predicted 95.58% of the variance in Cd data in BC-amended soils. The study highlights the significance of machine learning models that can predict the impact of BC amendments on soil's chemical and physical properties, such as pH, OC, clay minerals, and Cd availability. Machine learning models are capable of generating precise assessments of soil Cd content and predicting the probability of soil Cd transformation and immobilization. These data could establish sustainable soil management and environmental protection strategies. The study proposes the development of ecological soil remediation systems to efficiently address heavy metal contamination in soils, with the goal of promoting environmental sustainability.

Supplementary Materials: The following supporting information can be downloaded at: https://www.mdpi.com/article/10.3390/toxics12080535/s1, Table S1: Area description of the soil samples collected; Table S2: Rice straw biochar properties; Table S3. Comparison with different studies.

Author Contributions: M.S.R.: conceptualization, data curation, investigation, software, and writing—original draft. Y.W.: writing—review and editing. Y.Y.: writing—review and editing. B.Y.: writing—review and editing, Software. S.J.: writing—review and editing. A.F.M.: Formal analysis, Investigation, Software, writing—review and editing. B.C.: writing—review and editing. X.L.: Investigation, Resources, writing—review and editing. Z.L.: Conceptualization, Methodology, Supervision, Resources, writing—review and editing. All authors have read and agreed to the published version of the manuscript.

Funding: This work was jointly supported by the open competition program of the top ten critical priorities of Agricultural Science and Technology Innovation for the 14th Five-Year Plan of Guangdong Province (Grant No:2023SDZG08),the Low Carbon Agriculture and Carbon Neutralization Research Center, GDAAS (XTXM202204) and the High-level Guangdong Agricultural Science and Technology Demonstration City Construction Fund City Institute Cooperation Project (2220060000054, 2220060000050).

Institutional Review Board Statement: Not applicable.

Informed Consent Statement: Not applicable.

Data Availability Statement: The data presented in this study are available on request from the corresponding author. The data are not publicly available due to privacy.

Conflicts of Interest: The authors declare that they have no known competing financial interests or personal relationships that could have appeared to influence the work reported in this paper.

References

1. Nawab, J.; Ghani, J.; Khan, S.; Xiaoping, W. Minimizing the Risk to Human Health Due to the Ingestion of Arsenic and Toxic Metals in Vegetables by the Application of Biochar, Farmyard Manure and Peat Moss. *J. Environ. Manag.* **2018**, *214*, 172–183. [CrossRef] [PubMed]
2. Song, C.; Ye, F.; Zhang, H.; Hong, J.; Hua, C.; Wang, B.; Chen, Y.; Ji, R.; Zhao, L. Metal(Loid) Oxides and Metal Sulfides Nanomaterials Reduced Heavy Metals Uptake in Soil Cultivated Cucumber Plants. *Environ. Pollut.* **2019**, *255*, 113354. [CrossRef] [PubMed]
3. Guo, J.; Yan, C.; Luo, Z.; Fang, H.; Hu, S.; Cao, Y. Synthesis of a Novel Ternary HA/Fe-Mn Oxides-Loaded Biochar Composite and Its Application in Cadmium(II) and Arsenic(V) Adsorption. *J. Environ. Sci.* **2019**, *85*, 168–176. [CrossRef] [PubMed]
4. Liao, X.; Zhang, C.; Sun, G.; Li, Z.; Shang, L.; Fu, Y.; He, Y.; Yang, Y. Assessment of Metalloid and Metal Contamination in Soils from Hainan, China. *Int. J. Environ. Res. Public Health* **2018**, *15*, 454. [CrossRef] [PubMed]
5. Li, H.; Wu, W.; Min, X.; Zhan, W.; Fang, T.; Dong, X.; Shi, Y. Immobilization and Assessment of Heavy Metals in Chicken Manure Compost Amended with Rice Straw-Derived Biochar. *Environ. Pollut. Bioavailab.* **2021**, *33*, 1–10. [CrossRef]
6. Mahfooz, Y.; Yasar, A.; Guijian, L.; Islam, Q.U.; Akhtar, A.B.T.; Rasheed, R.; Irshad, S.; Naeem, U. Critical Risk Analysis of Metals Toxicity in Wastewater Irrigated Soil and Crops: A Study of a Semi-Arid Developing Region. *Sci. Rep.* **2020**, *10*, 1–11. [CrossRef]
7. Wu, D.; Shi, Y.; Wang, M.; Ran, M.; Wang, Y.; Tian, L.; Ye, H.; Han, F. A Baseline Study on the Distribution Characteristics and Health Risk Assessment of Cadmium in Edible Tissues of the Swimming Crabs (*Portunus trituberculatus*) from Shanghai, China. *Mar. Pollut. Bull.* **2022**, *185*, 114253. [CrossRef]
8. Jian, S.; Lei, Y.; Li, B.; Lv, Y.; Gao, X.; Yang, X. Investigation of Sulfate on the Migration/Solidification Mechanism and Control of Cadmium during Sintering in Lightweight Aggregates. *Constr. Build. Mater.* **2022**, *352*, 129041. [CrossRef]
9. Zhang, P.; Xue, B.; Jiao, L.; Meng, X.; Zhang, L.; Li, B.; Sun, H. Preparation of Ball-Milled Phosphorus-Loaded Biochar and Its Highly Effective Remediation for Cd- and Pb-Contaminated Alkaline Soil. *Sci. Total Environ.* **2022**, *813*, 152648. [CrossRef]
10. Munir, M.A.M.; Liu, G.; Yousaf, B.; Ali, M.U.; Cheema, A.I.; Rashid, M.S.; Rehman, A. Bamboo-Biochar and Hydrothermally Treated-Coal Mediated Geochemical Speciation, Transformation and Uptake of Cd, Cr, and Pb in a Polymetal(Iod)s-Contaminated Mine Soil. *Environ. Pollut.* **2020**, *265*, 114816. [CrossRef]
11. Lu, K.; Yang, X.; Gielen, G.; Bolan, N.; Ok, Y.S.; Niazi, N.K.; Xu, S.; Yuan, G.; Chen, X.; Zhang, X.; et al. Effect of Bamboo and Rice Straw Biochars on the Mobility and Redistribution of Heavy Metals (Cd, Cu, Pb and Zn) in Contaminated Soil. *J. Environ. Manag.* **2017**, *186*, 285–292. [CrossRef]
12. Wang, S.; Xu, Y.; Norbu, N.; Wang, Z. Remediation of Biochar on Heavy Metal Polluted Soils. *IOP Conf. Ser. Earth Environ. Sci.* **2018**, *108*, 042113. [CrossRef]
13. Meng, F.; Huang, Q.; Cai, Y.; Xiao, L.; Wang, T.; Li, X.; Wu, W.; Yuan, G. A Comparative Assessment of Humic Acid and Biochar Altering Cadmium and Arsenic Fractions in a Paddy Soil. *J. Soils Sediments* **2023**, *23*, 845–855. [CrossRef]
14. Lu, Y.; Cheng, J.; Wang, J.; Zhang, F.; Tian, Y.; Liu, C.; Cao, L.; Zhou, Y. Efficient Remediation of Cadmium Contamination in Soil by Functionalized Biochar: Recent Advances, Challenges, and Future Prospects. *Processes* **2022**, *10*, 1627. [CrossRef]
15. Yu, K.; Fang, S.; Zhao, Y. Heavy Metal Hg Stress Detection in Tobacco Plant Using Hyperspectral Sensing and Data-Driven Machine Learning Methods. *Spectrochim. Acta Part A Mol. Biomol. Spectrosc.* **2021**, *245*, 118917. [CrossRef] [PubMed]
16. Li, J.; Yang, Z.; Zhao, Y.; Yu, K. HSI Combined with CNN Model Detection of Heavy Metal Cu Stress Levels in Apple Rootstocks. *Microchem. J.* **2023**, *194*, 109306. [CrossRef]
17. Sun, J.; Cao, Y.; Zhou, X.; Wu, M.; Sun, Y.; Hu, Y. Detection for Lead Pollution Level of Lettuce Leaves Based on Deep Belief Network Combined with Hyperspectral Image Technology. *J. Food Saf.* **2021**, *41*, e12866. [CrossRef]

18. Lu, X.Q.; Tian, J.; Liao, Q.; Xu, Z.W.; Gan, L. CNN-LSTM Based Incremental Attention Mechanism Enabled Phase-Space Reconstruction for Chaotic Time Series Prediction. *J. Electron. Sci. Technol.* **2024**, *22*, 100256. [CrossRef]
19. Chopannejad, S.; Roshanpoor, A.; Sadoughi, F. Attention-Assisted Hybrid CNN-BILSTM-BiGRU Model with SMOTE–Tomek Method to Detect Cardiac Arrhythmia Based on 12-Lead Electrocardiogram Signals. *Digit. Health* **2024**, *10*. [CrossRef]
20. da Silva, D.G.; de Moura Meneses, A.A. Comparing Long Short-Term Memory (LSTM) and Bidirectional LSTM Deep Neural Networks for Power Consumption Prediction. *Energy Rep.* **2023**, *10*, 3315–3334. [CrossRef]
21. Nguyen, T.N.; Lee, S.; Nguyen-Xuan, H.; Lee, J. A Novel Analysis-Prediction Approach for Geometrically Nonlinear Problems Using Group Method of Data Handling. *Comput. Methods Appl. Mech. Eng.* **2019**, *354*, 506–526. [CrossRef]
22. Deng, Y.; Huang, S.; Laird, D.A.; Wang, X.; Meng, Z. Adsorption Behaviour and Mechanisms of Cadmium and Nickel on Rice Straw Biochars in Single- and Binary-Metal Systems. *Chemosphere* **2019**, *218*, 308–318. [CrossRef] [PubMed]
23. Rahim, H.U.; Akbar, W.A.; Alatalo, J.M. A Comprehensive Literature Review on Cadmium (Cd) Status in the Soil Environment and Its Immobilization by Biochar-Based Materials. *Agronomy* **2022**, *12*, 877. [CrossRef]
24. Li, D.; Lai, C.; Li, Y.; Li, H.; Chen, G.; Lu, Q. Biochar Improves Cd-Contaminated Soil and Lowers Cd Accumulation in Chinese Flowering Cabbage (*Brassica parachinensis* L.). *Soil Tillage Res.* **2021**, *213*, 105085. [CrossRef]
25. Zhang, H.; Yue, X.; Li, F.; Xiao, R.; Zhang, Y.; Gu, D. Preparation of Rice Straw-Derived Biochar for Efficient Cadmium Removal by Modification of Oxygen-Containing Functional Groups. *Sci. Total Environ.* **2018**, *631–632*, 795–802. [CrossRef]
26. Qi, F.; Lamb, D.; Naidu, R.; Bolan, N.S.; Yan, Y.; Ok, Y.S.; Rahman, M.M.; Choppala, G. Cadmium Solubility and Bioavailability in Soils Amended with Acidic and Neutral Biochar. *Sci. Total Environ.* **2018**, *610–611*, 1457–1466. [CrossRef] [PubMed]
27. Antonangelo, J.A.; Sun, X.; Zhang, H. The Roles of Co-Composted Biochar (COMBI) in Improving Soil Quality, Crop Productivity, and Toxic Metal Amelioration. *J. Environ. Manag.* **2021**, *277*, 111443. [CrossRef] [PubMed]
28. Chagas, J.K.M.; Figueiredo, C.C.d.; da Silva, J.; Paz-Ferreiro, J. The Residual Effect of Sewage Sludge Biochar on Soil Availability and Bioaccumulation of Heavy Metals: Evidence from a Three-Year Field Experiment. *J. Environ. Manag.* **2021**, *279*, 111824. [CrossRef]
29. Yuan, C.; Li, Q.; Sun, Z.; Sun, H. Effects of Natural Organic Matter on Cadmium Mobility in Paddy Soil: A Review. *J. Environ. Sci.* **2021**, *104*, 204–215. [CrossRef]
30. Zhang, J.; Li, S. Air Quality Index Forecast in Beijing Based on CNN-LSTM Multi-Model. *Chemosphere* **2022**, *308*, 136180. [CrossRef]
31. Mei, P.; Li, M.; Zhang, Q.; Li, G.; Song, L. Prediction Model of Drinking Water Source Quality with Potential Industrial-Agricultural Pollution Based on CNN-GRU-Attention. *J. Hydrol.* **2022**, *610*, 127934. [CrossRef]
32. Nguyen, H.; Vu, T.; Vo, T.P.; Thai, H.-T. Efficient Machine Learning Models for Prediction of Concrete Strengths. *Constr. Build. Mater.* **2021**, *266*, 120950. [CrossRef]
33. Iluore, K.; Lu, J. Long Short-Term Memory and Gated Recurrent Neural Networks to Predict the Ionospheric Vertical Total Electron Content. *Adv. Sp. Res.* **2022**, *70*, 652–665. [CrossRef]
34. Haris, M.; Hamid, Y.; Usman, M.; Wang, L.; Saleem, A.; Su, F.; Guo, J.K.; Li, Y. Crop-Residues Derived Biochar: Synthesis, Properties, Characterization and Application for the Removal of Trace Elements in Soils. *J. Hazard. Mater.* **2021**, *416*, 126212. [CrossRef]
35. Abbas, Q.; Liu, G.; Yousaf, B.; Ali, M.U.; Ullah, H.; Ahmed, R. Effects of Biochar on Uptake, Acquisition and Translocation of Silver Nanoparticles in Rice (*Oryza sativa* L.) in Relation to Growth, Photosynthetic Traits and Nutrients Displacement. *Environ. Pollut.* **2019**, *250*, 728–736. [CrossRef] [PubMed]
36. Walkley, A.J.; Black, I.A. Standard Operating Procedure for Soil Organic Carbon Walkley-Black Method. *Food Agric. Organ. United Nations* **1934**, *1*, 27.
37. Ross, D.; Kettering, Q. Recommended Methods for Determining Soil Cation Exchange Capacity. Chapter 9. In *Recommended Soil Testing Procedures for the Northeastern United States*; Bulletin No 193; College of Agriculture & Natural Resources: Newark, DE, USA, 2011; pp. 75–86.
38. Jin, X.; Zhang, J.; Wang, X.; Zhang, X.; Guo, T.; Shi, C.; Su, T.; Kong, J.; Bai, Y. A Deep Network Prediction Model for Heavy Metal Cadmium in the Rice Supply Chain. *J. Futur. Foods* **2021**, *1*, 196–202. [CrossRef]
39. Greeshma, M.; Simon, P. Bidirectional Gated Recurrent Unit with Glove Embedding and Attention Mechanism for Movie Review Classification. *Procedia Comput. Sci.* **2024**, *233*, 528–536. [CrossRef]
40. Alzubaidi, L.; Zhang, J.; Humaidi, A.J.; Al-Dujaili, A.; Duan, Y.; Al-Shamma, O.; Santamaría, J.; Fadhel, M.A.; Al-Amidie, M.; Farhan, L. Review of Deep Learning: Concepts, CNN Architectures, Challenges, Applications, Future Directions. *J. Big Data* **2021**, *8*, 53. [CrossRef]
41. El-Naggar, A.; Shaheen, S.M.; Ok, Y.S.; Rinklebe, J. Biochar Affects the Dissolved and Colloidal Concentrations of Cd, Cu, Ni, and Zn and Their Phytoavailability and Potential Mobility in a Mining Soil under Dynamic Redox-Conditions. *Sci. Total Environ.* **2018**, *624*, 1059–1071. [CrossRef]
42. Liu, L.; Huang, L.; Huang, R.; Lin, H.; Wang, D. Immobilization of Heavy Metals in Biochar Derived from Co-Pyrolysis of Sewage Sludge and Calcium Sulfate. *J. Hazard. Mater.* **2021**, *403*, 123648. [CrossRef] [PubMed]
43. Trakal, L.; Vítková, M.; Hudcová, B.; Beesley, L.; Komárek, M. Chapter 7—Biochar and Its Composites for Metal(Loid) Removal from Aqueous Solutions. In *Biochar from Biomass and Waste, Fundamentals and Applications*; Elsevier: Amsterdam, The Netherlands, 2018; pp. 113–141. [CrossRef]

44. Liao, S.; Pan, B.; Li, H.; Zhang, D.; Xing, B. Detecting Free Radicals in Biochars and Determining Their Ability to Inhibit the Germination and Growth of Corn, Wheat and Rice Seedlings. *Environ. Sci. Technol.* **2014**, *48*, 8581–8587. [CrossRef] [PubMed]
45. Lu, H.; Li, K.; Nkoh, J.N.; Shi, Y.; He, X.; Hong, Z.; Xu, R. Effects of the Increases in Soil PH and PH Buffering Capacity Induced by Crop Residue Biochars on Available Cd Contents in Acidic Paddy Soils. *Chemosphere* **2022**, *301*, 134674. [CrossRef] [PubMed]
46. Bogusz, A.; Oleszczuk, P. Effect of Biochar Addition to Sewage Sludge on Cadmium, Copper and Lead Speciation in Sewage Sludge-Amended Soil. *Chemosphere* **2020**, *239*, 124719. [CrossRef] [PubMed]
47. Quan, G.; Fan, Q.; Sun, J.; Cui, L.; Wang, H.; Gao, B.; Yan, J. Characteristics of Organo-Mineral Complexes in Contaminated Soils with Long-Term Biochar Application. *J. Hazard. Mater.* **2020**, *384*, 121265. [CrossRef]
48. Zhu, L.x.; Xiao, Q.; Shen, Y.f.; Li, S.Q. Effects of Biochar and Maize Straw on the Short-Term Carbon and Nitrogen Dynamics in a Cultivated Silty Loam in China. *Environ. Sci. Pollut. Res.* **2017**, *24*, 1019–1029. [CrossRef]
49. Ruan, X.; Sun, Y.; Du, W.; Tang, Y.; Liu, Q.; Zhang, Z.; Doherty, W.; Frost, R.L.; Qian, G.; Tsang, D.C.W.W. Formation, Characteristics, and Applications of Environmentally Persistent Free Radicals in Biochars: A Review. *Bioresour. Technol.* **2019**, *281*, 457–468. [CrossRef] [PubMed]
50. Han, L.; Sun, K.; Yang, Y.; Xia, X.; Li, F.; Yang, Z.; Xing, B. Biochar's Stability and Effect on the Content, Composition and Turnover of Soil Organic Carbon. *Geoderma* **2020**, *364*, 114184. [CrossRef]
51. Qi, F.; Yan, Y.; Lamb, D.; Naidu, R.; Bolan, N.S.; Liu, Y.; Ok, Y.S.; Donne, S.W.; Semple, K.T. Thermal Stability of Biochar and Its Effects on Cadmium Sorption Capacity. *Bioresour. Technol.* **2017**, *246*, 48–56. [CrossRef]
52. Sun, J.; Fan, Q.; Ma, J.; Cui, L.; Quan, G.; Yan, J.; Wu, L.; Hina, K.; Abdul, B.; Wang, H. Effects of Biochar on Cadmium (Cd) Uptake in Vegetables and Its Natural Downward Movement in Saline-Alkali Soil. *Environ. Pollut. Bioavailab.* **2020**, *32*, 36–46. [CrossRef]
53. Jing, F.; Sun, Y.; Liu, Y.; Wan, Z.; Chen, J.; Tsang, D.C.W. Interactions between Biochar and Clay Minerals in Changing Biochar Carbon Stability. *Sci. Total Environ.* **2022**, *809*, 151124. [CrossRef] [PubMed]
54. Lin, Y.; Munroe, P.; Joseph, S.; Ziolkowski, A.; van Zwieten, L.; Kimber, S.; Rust, J. Chemical and Structural Analysis of Enhanced Biochars: Thermally Treated Mixtures of Biochar, Chicken Litter, Clay and Minerals. *Chemosphere* **2013**, *91*, 35–40. [CrossRef] [PubMed]
55. Li, L.-P.; Liu, Y.-H.; Ren, D.; Wang, J.-J. Characteristics and Chlorine Reactivity of Biochar-Derived Dissolved Organic Matter: Effects of Feedstock Type and Pyrolysis Temperature. *Water Res.* **2022**, *211*, 118044. [CrossRef]
56. Jiang, X.; Tan, X.; Cheng, J.; Haddix, M.L.; Cotrufo, M.F. Interactions between Aged Biochar, Fresh Low Molecular Weight Carbon and Soil Organic Carbon after 3.5 years Soil-Biochar Incubations. *Geoderma* **2019**, *333*, 99–107. [CrossRef]
57. Xue, B.; Huang, L.; Li, X.; Lu, J.; Gao, R.; Kamran, M.; Fahad, S. Effect of Clay Mineralogy and Soil Organic Carbon in Aggregates under Straw Incorporation. *Agronomy* **2022**, *12*, 534. [CrossRef]
58. Xue, S.; Chen, F.; Wang, Y.; Shao, Z.; Zhang, C.; Qiu, L.; Ran, Y.; He, L. Effects of Co-Applications of Biochar and Solid Digestate on Enzyme Activities and Heavy Metals Bioavailability in Cd-Polluted Greenhouse Soil. *Water, Air, Soil Pollut.* **2021**, *232*, 140. [CrossRef]
59. Yan, Z.; Ding, W.; Xie, G.; Yan, M.; Li, J.; Han, Y.; Xiong, X.; Wang, C. Identification of Cadmium Phytoavailability in Response to Cadmium Transformation and Changes in Soil PH and Electrical Conductivity. *Chemosphere* **2023**, *342*, 140042. [CrossRef]
60. Ullah, A.; Ren, W.-L.; Tian, P.; Yu, X.-Z. Biochar as a Green Strategy in Alleviating Cd Mobility in Soil and Uptake in Plants: A Step towards Cd-Free Food. *Int. Biodeterior. Biodegradation* **2024**, *190*, 105787. [CrossRef]
61. Jing, F.; Chen, C.; Chen, X.; Liu, W.; Wen, X.; Hu, S.; Yang, Z.; Guo, B.; Xu, Y.; Yu, Q. Effects of Wheat Straw Derived Biochar on Cadmium Availability in a Paddy Soil and Its Accumulation in Rice. *Environ. Pollut.* **2020**, *257*, 113592. [CrossRef]
62. Nguyen, T.-B.; Sherpa, K.; Bui, X.-T.; Nguyen, V.-T.; Vo, T.-D.-H.; Ho, H.-T.-T.; Chen, C.-W.; Dong, C.-D. Biochar for Soil Remediation: A Comprehensive Review of Current Research on Pollutant Removal. *Environ. Pollut.* **2023**, *337*, 122571. [CrossRef]
63. Ghassemi-Golezani, K.; Farhangi-Abriz, S. Biochar Related Treatments Improved Physiological Performance, Growth and Productivity of *Mentha crispa* L. Plants under Fluoride and Cadmium Toxicities. *Ind. Crops Prod.* **2023**, *194*, 116287. [CrossRef]
64. Wei, B.; Zhang, D.; Jeyakumar, P.; Trakal, L.; Wang, H.; Sun, K.; Wei, Y.; Zhang, X.; Ling, H.; He, S.; et al. Iron-Modified Biochar Effectively Mitigates Arsenic-Cadmium Pollution in Paddy Fields: A Meta-Analysis. *J. Hazard. Mater.* **2024**, *469*, 133866. [CrossRef] [PubMed]
65. Zhao, X.-Y.; Zhang, Z.-Y.; Huang, Y.-M.; Feng, F.-J. Enhancing the Effect of Biochar Ageing on Reducing Cadmium Accumulation in *Medicago sativa* L. *Sci. Total Environ.* **2023**, *862*, 160690. [CrossRef] [PubMed]
66. Wei, B.; Peng, Y.; Jeyakumar, P.; Lin, L.; Zhang, D.; Yang, M.; Zhu, J.; Ki Lin, C.S.; Wang, H.; Wang, Z.; et al. Soil PH Restricts the Ability of Biochar to Passivate Cadmium: A Meta-Analysis. *Environ. Res.* **2023**, *219*, 115110. [CrossRef] [PubMed]
67. Li, G.; Chen, F.; Jia, S.; Wang, Z.; Zuo, Q.; He, H. Effect of Biochar on Cd and Pyrene Removal and Bacteria Communities Variations in Soils with Culturing Ryegrass (*Lolium perenne* L.). *Environ. Pollut.* **2020**, *265*, 114887. [CrossRef] [PubMed]
68. Palansooriya, K.N.; Li, J.; Dissanayake, P.D.; Suvarna, M.; Li, L.; Yuan, X.; Sarkar, B.; Tsang, D.C.W.; Rinklebe, J.; Wang, X.; et al. Prediction of Soil Heavy Metal Immobilization by Biochar Using Machine Learning. *Environ. Sci. Technol.* **2022**, *56*, 4187–4198. [CrossRef] [PubMed]
69. Song, P.; Liu, J.; Ma, W.; Gao, X. Remediation of Cadmium-Contaminated Soil by Biochar-Loaded Nano-Zero-Valent Iron and Its Microbial Community Responses. *J. Environ. Chem. Eng.* **2024**, *12*, 112311. [CrossRef]

70. Liu, J.; He, T.; Yang, Z.; Peng, S.; Zhu, Y.; Li, H.; Lu, D.; Li, Q.; Feng, Y.; Chen, K.; et al. Insight into the Mechanism of Nano-TiO2-Doped Biochar in Mitigating Cadmium Mobility in Soil-Pak Choi System. *Sci. Total Environ.* **2024**, *916*, 169996. [CrossRef]
71. Huang, Y.; Liu, T.; Liu, J.; Xiao, X.; Wan, Y.; An, H.; Luo, X.; Luo, S. Exceptional Anti-Toxic Growth of Water Spinach in Arsenic and Cadmium Co-Contaminated Soil Remediated Using Biochar Loaded with Bacillus Aryabhattai. *J. Hazard. Mater.* **2024**, *469*, 133966. [CrossRef]
72. Xu, W.; Xie, X.; Li, Q.; Yang, X.; Ren, J.; Shi, Y.; Liu, D.; Shaheen, S.M.; Rinklebe, J. Biochar Co-Pyrolyzed from Peanut Shells and Maize Straw Improved Soil Biochemical Properties, Rice Yield, and Reduced Cadmium Mobilization and Accumulation by Rice: Biogeochemical Investigations. *J. Hazard. Mater.* **2024**, *466*, 133486. [CrossRef]
73. Arif, M.; Liu, G.; Zia ur Rehman, M.; Yousaf, B.; Ahmed, R.; Mian, M.M.; Ashraf, A.; Mujtaba Munir, M.A.; Rashid, M.S.; Naeem, A. Carbon Dioxide Activated Biochar-Clay Mineral Composite Efficiently Removes Ciprofloxacin from Contaminated Water—Reveals an Incubation Study. *J. Clean. Prod.* **2022**, *332*, 130079. [CrossRef]
74. Arif, M.; Liu, G.; Yousaf, B.; Ahmed, R.; Irshad, S.; Ashraf, A.; Zia-ur-Rehman, M.; Rashid, M.S. Synthesis, Characteristics and Mechanistic Insight into the Clays and Clay Minerals-Biochar Surface Interactions for Contaminants Removal-A Review. *J. Clean. Prod.* **2021**, *310*, 127548. [CrossRef]
75. Palansooriya, K.N.; Shaheen, S.M.; Chen, S.S.; Tsang, D.C.W.; Hashimoto, Y.; Hou, D.; Bolan, N.S.; Rinklebe, J.; Ok, Y.S. Soil Amendments for Immobilization of Potentially Toxic Elements in Contaminated Soils: A Critical Review. *Environ. Int.* **2020**, *134*, 105046. [CrossRef] [PubMed]
76. Fang, Y.; Wang, P.; Zhang, L.; Zhang, H.; Xiao, R.; Luo, Y.; Tang, K.H.D.; Li, R.; Abdelrahman, H.; Zhang, Z.; et al. A Novel Zr-P-Modified Nanomagnetic Herbal Biochar Immobilized Cd and Pb in Water and Soil and Enhanced the Relative Abundance of Metal-Resistant Bacteria: Biogeochemical and Spectroscopic Investigations to Identify the Governing Factors and Potential Mech. *Chem. Eng. J.* **2024**, *485*, 149978. [CrossRef]
77. Mujtaba Munir, M.A.; Liu, G.; Yousaf, B.; Ali, M.U.; Abbas, Q.; Ullah, H. Synergistic Effects of Biochar and Processed Fly Ash on Bioavailability, Transformation and Accumulation of Heavy Metals by Maize (*Zea mays* L.) in Coal-Mining Contaminated Soil. *Chemosphere* **2020**, *240*, 124845. [CrossRef]

toxics

MDPI

Article

Rice Husk as a Sustainable Amendment for Heavy Metal Immobilization in Contaminated Soils: A Pathway to Environmental Remediation

Riccardo Cecire [1,2], Aleandro Diana [1,*], Agnese Giacomino [3], Ornella Abollino [3], Paolo Inaudi [3], Laura Favilli [3], Stefano Bertinetti [1], Simone Cavalera [1], Luisella Celi [2] and Mery Malandrino [1,*]

[1] Department of Chemistry, University of Turin, Via Giuria 7, 10125 Turin, Italy; riccardo.cecire@unito.it (R.C.); stefano.bertinetti@unito.it (S.B.); simone.cavalera@unito.it (S.C.)

[2] Department of Agricultural, Forest and Food Sciences, University of Turin, Largo Paolo Braccini 2, 10095 Grugliasco, Italy; luisella.celi@unito.it

[3] Department of Drug Science and Technology, University of Turin, Via Giuria 9, 10125 Turin, Italy; agnese.giacomino@unito.it (A.G.); ornella.abollino@unito.it (O.A.); paolo.inaudi@unito.it (P.I.); laura.favilli@unito.it (L.F.)

* Correspondence: aleandro.diana@unito.it (A.D.); mery.malandrino@unito.it (M.M.)

Abstract: Rice husk is a waste byproduct of rice production. This material has a moderate cost and is readily available, representing 20–22% of the biomass produced by rice cultivation. This study focused on the properties of rice husk in the remediation of soils contaminated by heavy metals. The effect of particle size, pH, and the presence of organic ligands on sorption efficiency was evaluated for Cd, Cu, and Mn. The continuous flow method was used to select suitable operative conditions and maximize the retention of heavy metals. Subsequently, pot experiments were carried out by growing two broadleaf plants, *Lactuca sativa* and *Spinacia oleracea*, in aliquots of soil collected in a Piedmont (Northwest Italy) site heavily contaminated by Cu, Cr, and Ni. Rice husk was added to the contaminated soil to evaluate its effectiveness in immobilizing heavy metals. The availability of Cr, Mn, Ni, Cu, Zn, Cd, and Pb in soil was studied using Tessier's sequential extraction protocol. The content of the elements was also analyzed in plants and the uptake of heavy metals was evaluated in relation to the addition of rice husk. The growth of both plants was more efficient in the presence of rice husk due to its ability to reduce the mobility of heavy metals in the soil. The simplicity, cost-effectiveness, and scalability of its employment make the use of rice husk suitable for practical applications in soil remediation.

Keywords: soil remediation; soil amendment; broadleaf plants; potentially toxic elements; Tessier protocol; pot experiments

1. Introduction

Food security is a critical issue for peri-urban areas, as they are vulnerable to contamination due to the coexistence of various anthropogenic activities, including industrial discharges, agricultural runoff, and atmospheric deposition [1–3]. The presence of heavy metals in peri-urban soils is of great concern because of their adverse effects on the environment, human health, and agriculture [4,5].

Plants can be overly sensitive to heavy metals, which can trigger various physiological responses even at low stress levels. While some heavy metals are essential for plant growth, high concentrations can cause toxicity, compromising physiological processes and posing a risk to food safety [6]. The effects of heavy metal contamination on plants are complex and depend on the specific metal; for instance, Cd and Pb affect respiration and photosynthesis, while Zn and Cu disrupt cell function and membrane integrity [6]. The mobility and availability of heavy metals to plants are contingent upon soil characteristics, including mineral composition, organic matter content, pH, and redox potential [2].

Heavy metals, in contrast to organic pollutants, exhibit remarkable persistence in soils as they resist microbial or chemical degradation [3]. Increases in public awareness about the health consequences of contaminated soils have promoted the creation of innovative approaches to their remediation, especially considering the prohibitive costs associated with traditional methods [1].

Selecting the most suitable remediation technology hinges on several critical factors, including the type and extent of contamination, the economic feasibility, and the overall efficiency of the chosen method. In peri-urban areas, the complexity of contamination sources and land use patterns necessitates a careful evaluation of these factors to determine the optimal approach. Efficiency is a paramount consideration when choosing remediation technology, and it is essential to assess whether the method can effectively reduce the bioavailability of heavy metals in the soil, thus minimizing risks to human health and the environment. Economic feasibility is another critical factor, and the cost of implementing remediation technology, including material acquisition, pre-treatment cost, labor, and maintenance, must be balanced against the potential benefits and long-term impacts [7].

The remediation of contaminated soils involves a range of approaches that can be broadly categorized into physical, chemical, and biological methods [8,9]. Each category has its unique advantages and limitations, making the selection of the most appropriate technique contingent on the specific characteristics of the contamination and the environmental context. Chemical methods, which encompass various techniques for altering the chemical properties of the soil, have shown promising results in immobilizing contaminants and reducing their bioavailability [10,11]. In order to achieve this objective, various soil amendments, such as phosphate compounds, liming materials, organic matter residues, metal oxides, and biochar, are currently studied as options for soil remediation [12]. These amendments interact with heavy metals in the soil and transform them into less mobile and less bioavailable forms.

Rice husk is an environmentally friendly and readily available material, representing 20–22% of the biomass produced by rice cultivation. It is rich in silica (20%), which has a high affinity for heavy metals due to the existence of ionizable silanol groups. Furthermore, rice husk contains hemicellulose (12%) and proteins (3%) that, with carboxyl and amine groups, can give interactions with heavy metals. Rice husks also contain organic acids such as oxalic, citric, acetic, malic, and succinic, which can form stable complexes with heavy metals [13]. All these components in rice husk can facilitate the adsorption and complexation of heavy metals in soil, reducing their mobility and bioavailability. The low management costs and the fact that it is a waste product make it a cost-effective solution for soil remediation. Additionally, utilizing rice husk for soil remediation exemplifies the principles of waste upscaling and valorization, transforming an agricultural byproduct into a valuable resource. This approach not only mitigates waste disposal issues but also provides an innovative and sustainable method for addressing soil contamination [14].

Furthermore, rice husk-derived materials have garnered attention for their remarkable adsorption properties and their potential for immobilizing heavy metals in wastewaters and soil [1,15,16]. For instance, Zheng et al. (2012) explored the use of biochar derived from rice residue to assess its impact on heavy metal accumulation in rice seedlings [17]. It was found that while biochar effectively reduced the uptake of Cd, Zn, and Pb by rice plants, it increased the accumulation of As. This dual effect underscores the complexity of biochar application in agricultural settings. In addition, Derakhshan Nejad and Jung (2017) focused on rice husk biochar and its efficacy in immobilizing multiple heavy metals (Cu, Cd, Zn, Pb) in soil, highlighting rice husk biochar as a promising soil amendment for reducing heavy metals' phytoavailability and leachability [18]. Furthermore, Bian et al. (2022) conducted pot experiments to evaluate the potential of rice husk-derived biochar to mitigate Cd uptake by cabbage leaves. Results indicated that biochar application significantly decreased Cd uptake, demonstrating its potential to enhance food safety and soil health [19].

Despite the substantial research focused on treated rice husk materials that show promising results in heavy metal remediation, there remains a notable gap in understanding the potential efficacy of untreated rice husk in this context. This underscores the necessity to explore the inherent properties of untreated rice husk without the additional costs associated with treatment processes, particularly in scaling up its application for soil remediation purposes.

In the present work, we have delved into the scientific principles behind rice husk's effectiveness as a soil amendment, its practical application, and the feasibility of implementing this approach in peri-urban areas.

This paper will discuss the characterization and feasibility of rice husk as an adsorbent of heavy metals (namely Cd, Cu, and Mn) as a function of the effect of particle size, pH, and the presence of organic ligands. Subsequently, the changes in metal availability in the soil after application of rice husk were studied. Finally, its application as a soil amendment in a heavy metal-contaminated soil sample was evaluated as well as its effectiveness in reducing the uptake of the pollutants by two plants, *Lactuca sativa* and *Spinacia oleracea*.

In the present work, the aim was to evaluate the potential of untreated rice husk as an amendment to reduce the bioavailability of heavy metals in contaminated soils. This was achieved by (1) characterizing the sorption behavior of rice husk towards heavy metals (Cd, Cu, Mn) based on particle size, pH, and the presence of organic ligands; (2) studying the changes in metal availability in soil after rice husk application; and (3) evaluating its effect on reducing heavy metal uptake by *Lactuca sativa* and *Spinacia oleracea*. The hypothesis is that untreated rice husk can effectively decrease the availability of heavy metals in soil because of its abundant silica and organic compounds content, offering an affordable and eco-friendly method for remediation.

2. Materials and Methods

2.1. Apparatus and Reagents

For acid digestion, an MLS-1200 MEGA microwave (Milestone, Sorisole, Italy) oven was used, equipped with a sample carousel with Teflon vessels fitted with a safety valve closure.

The concentration of the analytes in the tested solutions was determined by inductively coupled plasma atomic emission spectroscopy (ICP-AES) using a VARIAN LIBERTY 100 spectrometer (Varian Australia, Mullgrave, Australia). The wavelengths of the analytes chosen and the limits of detection (LOD) are present in Table S1.

Due to their lower concentration, Cd and Pb were also determined by graphite furnace atomic absorption spectrometry (GF-AAS) with a Perkin Elmer 5100 instrument (Perkin Elmer, Norwalk, CT, USA).

Determinations of the metals were performed using external calibration. The calibrations were always performed with standard solutions prepared in aliquots of sample blanks (matrix matching method). Metal ion standards were prepared from commercial 1000 mg/L stock solutions (MERCK Titrisol, Darmstadt, Germany; vials containing 1.000 ± 0.002 g metal in salt form). The reagents used are all analytical purity grade. All solutions were prepared using HPW (High Purity Water). Water was purified in a Milli-Q system, resulting in water with a resistivity of 18 MΩ·cm.

A pH meter, model EA 920, from the company ORION (Orion Research INC, Ayer, MA, USA) was used to measure the pH of solutions, equipped with a glass/calomel combination electrode. The instrument was calibrated daily using two standard solutions, one at pH 4 (citrate-HCl) and one at pH 7 (NaH_2PO_4-Na_2HPO_4). The determination is subject to an uncertainty of 0.02 pH units [20].

In Figure 1, a block diagram is illustrated that shows the experimental workflow of this study.

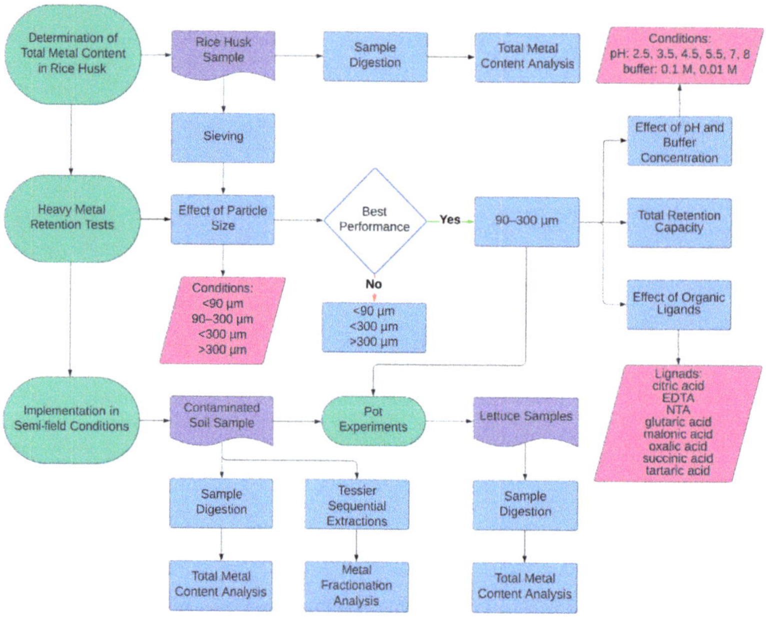

Figure 1. Block diagram of experiments performed.

2.2. Determination of Total Metal Content in Rice Husk

Rice husk was obtained from a farm in the province of Pavia (Lombardy, Northwest Italy). The inorganic content of the rice husk was characterized to assess the possible release of heavy metals during retention tests and to evaluate the possible input of contaminants into the soil. Three sample aliquots of 0.1 g of untreated rice husk were digested with 5 mL of aqua regia and 2 mL of hydrofluoric acid (HF). The following heating program was applied in the MW oven: four heating steps of 5 min each (250, 400, 600, 250 W, respectively). Subsequently, 0.7 g of boric acid (H_3BO_3) were added, and the samples were heated again for 5 min at 250 W. The resulting solutions were filtered with paper filters and diluted to 100 mL with HPW. The solutions were analyzed by ICP-AES.

2.3. Heavy Metal Retention Tests

A continuous flow system was employed (Figure 2) in preference of the batch technique to understand the effect of various considered parameters (particle size, pH, organic ligands) on the retention of metals by the rice husk, because this system can be easily scaled up from a laboratory system to a real application [21,22]. This system was preferred to better simulate a dynamic equilibrium condition of a soil subject to frequent environmental stresses than a static equilibrium situation.

Metal adsorption tests were conducted in polypropylene columns (Bio-Rad, Hercules, CA, USA; length 4 cm, internal diameter 5 mm) packed with 0.25 g of rice husk. At the bottom of the column, a cellulose acetate filter (Millipore Merck, Darmstadt, Germany) was placed on the porous polymer support to prevent clogging of the porous septum and the loss of the smallest rice husk particles into the eluates. The slurry packing method was used to obtain a homogeneous bed of particles in the column. The method consists of placing rice husk and 5 mL of HPW in a beaker to create a suspension.

Standard solutions of the metals under investigation were injected through the columns with the help of a Gilson Miniplus 2 (Gilson Middleton, CT, USA) multichannel peristaltic pump at a constant flow rate (0.5 mL/min). The pump was connected to the columns with polypropylene connectors and PVC tubing (i.e., 1.29 mm). Before each experiment, rice husk was conditioned to the working pH by flowing the proper buffer solution. Solutions of the metal ions at a concentration of 1×10^{-3} M were prepared and diluted with buffer solutions at the desired pH. The following buffer solutions were utilized: trichloroacetic acid (TCA) buffer, utilized for experiments with pH values less than 3; acetic acid/sodium acetate buffer (HAc/NaAc), utilized for experiments conducted at pH 3.3–5.5; HEPES buffer, utilized for pH 7 and 8 experiments.

Column retention experiments were conducted on untreated rice husk monitoring three metal ions: Cu, Cd, and Mn. Three replicates of each experiment were performed. For each experiment, a procedural blank was also performed to account for contamination or release of the metals from the rice husk. Finally, 25 mL of the eluate were collected and analyzed by ICP-AES. The amount of metal retained on the rice husk was calculated by the difference between the metal content in the input and output solution, corrected for the value of the blank, and expressed as a percentage.

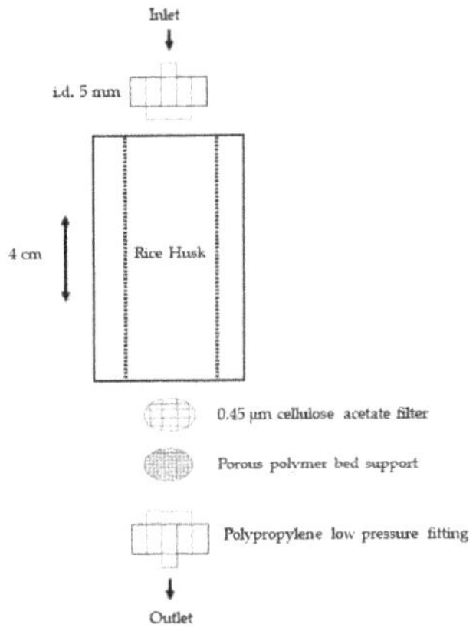

Figure 2. A schematic representation of the continuous flow system apparatus used for the experiments.

2.3.1. Effect of Particle Size

Rice husk contains particles of assorted sizes, ranging from small to coarse. Since it must be packed into columns, it is essential to have a sort of size uniformity, avoiding preferential paths. Therefore, the choice of particle size was crucial to improve the interaction with metal ions. Tests were conducted at pH 5.5 with a concentration of 0.1 M acetate buffer, 1×10^{-4} M metal ions (Cd, Cu, and Mn), and four particle size fractions of rice husk obtained after sieving: <90 µm; between 90 and 300 µm; <300 µm; >300 µm. The behavior of the metals was studied both individually and in mixtures.

2.3.2. Effect of pH and Buffer Concentration

Metal adsorption depends on numerous factors including pH and ionic strength. Industrial and processing waters and wastewaters, as well as contaminated soils, show often a broad range of pH values and behave differently toward divalent cations in competition with one another [23]. For these reasons, the effect of pH (2.5–8) and the buffer concentration (1×10^{-1} M and 1×10^{-2} M) on the retention of Cd, Cu, and Mn by rice husk was investigated. Experiments were conducted with metals concentration of 1×10^{-4} M in mixtures.

2.3.3. Effect of Ligands

The retention of metals by rice husk can be affected by the presence of ligands, contingent upon the properties of the ligand, rice husk, and metal under consideration. In this study, we studied the effect of following organic ligands: citric acid, ethylenediaminetetraacetic acid (EDTA), nitrilotriacetic acid (NTA), glutaric acid, malonic acid, oxalic acid, succinic acid, and tartaric acid. These are some of the most used ligands for metal complexation and their functional groups were used to simulate the behavior of more complex organic molecules potentially present in soils. The metal ion concentration was 1×10^{-4} M, and the ligand concentration was 3×10^{-4} M. The effect of the presence of ligands on the retention of the three metals was conducted at pH 3.5 with a buffer concentration of 1×10^{-2} M. It was decided to simulate an extreme environmental condition with a low pH to better simulate the acidic effluents characterizing pollution of our case study.

To analyze the data obtained in the experiment on the presence of ligands, One-Way ANOVA by the XLSTAT software (version 2022.2.1) was used to check if there was a statistical difference between the mean concentrations in the effluent solutions of the nine groups for each metal investigated. The assumption of normality and homogeneity of variances was assessed prior to conducting ANOVA to ensure the appropriateness of the test [24]. Following the ANOVA, a Dunnett post hoc test was applied to compare each treatment group against a single control group, determining whether significant differences existed between the control and the treated groups. This approach helps the control for Type I errors associated with multiple comparisons, ensuring robust statistical conclusions [25].

PyES (version 1.1.3), an open-access software that allows for the calculation of the distribution of species in solution under the experimental conditions, was used [26].

2.3.4. Total Retention Capacity

The total retention capacity of rice husk towards Cd, Cu, and Mn was studied in conditions of nearly quantitative uptake for all the metals to ensure that the amount of metal retained by the rice husk depends mainly on thermodynamic reasons. The saturation point was determined when the concentration of the influent solution (C_0) equaled that of the effluent solution (C), indicating full occupation of the rice husk's surface adsorption sites. The breakthrough curve was constructed by plotting the ratio of C/C_0 as a function of the effluent volume [27]. The set-up is the same as described in the previous paragraph except for the type of columns used: ECONO-Column Chromatography Bio-Rad, diameter 1.5 cm, volume 18 ml (Bio-Rad Laboratories, Hercules, CA, USA). Solutions of the metal ions at a concentration of 1×10^{-3} M were prepared and diluted with acetate buffer in concentration of 1×10^{-2} M at pH 5.5.

2.4. *Implementation in Semi-Field Conditions*

We tested rice husk as a soil amendment for the remediation of contaminated soil. Contaminated soil samples were derived from the town of Borgomanero, in the province of Novara (Piedmont, Northwest, Italy). A map of the location is presented in Figure S1 in the Supplementary Material. The area was previously used as a permanent meadow and woodland in the past and became contaminated because of the repeated floods of a small stream collecting the wastewater of local industries, some of which operate in the electroplating processes. Its floods caused an accumulation of inorganic contaminants in the soil. The core of the contaminated zone was about 3000 m^2 wide.

The soil sampling was originally conducted in 2012, following the work by Malandrino et al. (2011) [28]. Several sub-samples were collected across the contaminated site to ensure comprehensive coverage. These samples were properly stocked and preserved until their use in this experiment. For the current study, a composite soil sample was created by mixing the sub-samples collected during the 2012 sampling, ensuring homogeneity and representativeness of the contaminated site. This approach allowed for a consistent and well-characterized sample throughout the experiment, enhancing the reliability of the findings.

For the characteristics of the contaminated soil, see Table S2 and Malandrino et al. (2011) [28]. The contaminated soil texture is classified as loamy sand, and the soil pH is strongly acid (5.1–5.5) according to the USDA (2024) classification [29].

A control soil was also used. As shown in Table S2, the control soil was chosen based on its high organic matter content, substantial cation exchange capacity, and balanced texture to provide an ideal environment for the growth and development of lettuce and spinach, making it a suitable control for this experiment.

2.4.1. Determination of Total Metal Content in Soil

Soil samples were collected with plastic tools and transferred into polyethylene bags. Subsequently, each sample was air-dried, sieved through a 2 mm sieve, ground in a centrifugal ball mill, and stored in plastic bags prior to laboratory analysis. Three soil sample aliquots of 0.5 g were treated with a mixture of 10 mL of aqua regia and 4 mL of HF in PTFE vessels and mineralized in a microwave oven. Four heating steps of 5 min each (250, 400, 600, and 250 W respectively) were used. Then, 1.4 g of H_3BO_3 was added, and the vessels were further heated for 5 min at 250 W and again cooled by ventilation for 25 min. The resulting solutions were filtered with paper filters and diluted to 100 mL with HPW. The solutions were analyzed by ICP-AES and GF-AAS.

2.4.2. Tessier Fractionation

We evaluated how the addition of rice husk affected the mobility and reactivity of metals in soil using Tessier fractionation. This extraction procedure, adapted from Tessier et al. (1979), categorizes metals into five distinct chemical fractions: extractable and exchangeable, carbonate-bound, Fe and Mn oxide-bound, organic matter- and sulfide-bound, and residual [30]. The residual fraction is excluded as it is mainly integrated within the crystal structures of rocks and minerals, releasing only over the long term [31,32]. After each extraction, the suspension was centrifuged for 20 min at 4000 rpm. The supernatant was separated, while the solid residue was washed with 10 mL HPW and centrifuged again for 5 min. The washed water was added to the supernatant, while the solid residue was ready for the next extraction. The extracts were diluted to 25 mL (first fraction), 50 mL (second fraction), and 100 mL (third, fourth, and fifth fractions), stabilized by the addition of 25, 50, and 100 µL of concentrated HNO_3, respectively, and analyzed. The concentrations of Cr, Cu, Mn, Ni, and Zn in the five fractions were determined by ICP-AES, whereas Cd and Pb were determined by GF-AAS.

2.4.3. Effect on Lettuce and Spinach Heavy Metal Uptake

To evaluate the effectiveness of the rice husk in reducing the phytoavailability of the metal pollutants in a real scenario, we evaluated the effect of the addition of this material by measuring metal uptake by lettuce (*Lactuca sativa*) and spinach (*Spinacia oleracea*), used as test crops [33,34].

Pot experiments were made by putting untreated contaminated soil (5 kg) into polyethylene pots and adding rice husk (500 g). At the same time, an aliquot of the contaminated soil and one of unpolluted soil were left unamended and used as references. Each treatment was performed in triplicate. The experiment was conducted with two plant species, *Lactuca sativa* and *Spinacia oleracea*. The pots were laid out at room temperature (25 °C), and they were watered three times a week with 500 mL of tap water. Plants grown in the pots were harvested after 2 months.

All harvested plants were stripped of their roots, washed with HPW, oven-dried at 60 °C for 16 h, and then ground in an agate mortar. Then, 0.2 g of ground plant material was digested with 10 mL of concentrated HNO_3 in a MW oven, using four heating steps of 5 min each (250, 400, 600, and 250 W, respectively). After cooling, the digestion solutions were filtered with paper filters and diluted to 50 mL. The resulting solutions were analyzed by ICP-AES for the determination of Cd, Cr, Cu, Mn, Ni, Pb, and Zn.

3. Results and Discussion

3.1. Rice Husk Inorganic Characterization

Results obtained for the inorganic characterization of rice husk are reported in Table S3, expressed as mg/kg of rice husk. If all the inorganic components are considered, namely as ashes, it is up to 13.6% of the total mass. This is confirmed by Soltani et al. (2015), who found a value in the range of 15–20% and Bao (2023) in the range of 13–21% [35,36]. Concentrations are converted into their respective oxide and are compared with other studies in Table 1. The most abundant oxide is SiO_2, constituting 88% of the rice husk ashes, followed by K_2O (4.5%), CaO (2.1%), MgO (2.0%), Fe_2O_3 (1.5%), Al_2O_3 (1.3%), and Na_2O (0.4%). The result for silica is consistent with other works that attest a value in the range of 67–97%. The presence of silica and consequently the existence of easily deprotonated silanol groups in rice husk is the main cause of the retention capacity towards heavy metals. As far as the other constituents are considered, the order of abundance is variable and could reflect the different rice families and crops specific to the regions considered. MnO_2 and CuO were not determined in other studies, but in this work, they were found to be present in traces and count for 0.29% and 0.01% of the total ashes.

3.2. Heavy Metal Retention Tests

3.2.1. Effect of Particle Size

Results of the effect of particle size on the retention of Cd, Cu, and Mn are shown in Figure 3. In the tests with individual metal solutions, Cu was retained with a value of 100% in all particle size fractions. Cd and Mn present a maximum retention in the fraction between 90 and 300, respectively, of 100% and 88%. A particle size fraction below 90 μm showed a 67% retention for Cd and 54% for Mn, whereas the fraction below 300 μm retained about 87% of Cd and 76% of Mn. The particle size fraction above 300 μm showed the lowest percent of retention, less than 50%, for both metals. Mn always shows the lowest % of retention in all particle size fractions. It can be concluded from this experiment that rice husk has a higher affinity for Cu, followed by Cd and Mn. Moreover, it is evident that the fraction with a grain size between 90 and 300 μm has the highest retention values.

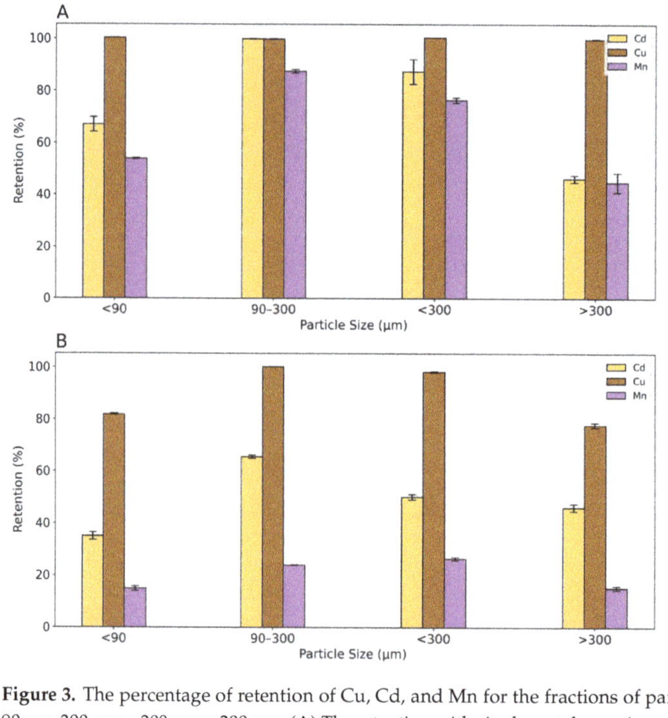

Figure 3. The percentage of retention of Cu, Cd, and Mn for the fractions of particle size: <90 μm; 90 μm–300 μm; <300 μm; >300 μm. (**A**) The retention with single metal experiments; (**B**) the retention with mixture metal experiments. Black bars represent confidence interval (95%).

Table 1. Chemical composition in percentage (%) of rice husk ashes from various countries expressed as oxides.

	Vietnam [1]	U.S.A. [1]	Thailand [1]	Nigeria [1]	North Ireland [1]	Malaysia [1]	Japan [1]	Iraq [1]	India [1]	Guyana [1]	Canada [1]	Brazil [1]	Italy [2]
SiO_2	86.9	87–97	89–95	67–76	86–96	93.1	91.6	86.8	86–94	88–95	87–97	92.9	**87.8**
Al_2O_3	0.84	Traces	0.5–1.0	3–4.90	0.08–0.84	0.21	0.14	0.4	0.2–5.0	-	0.15–0.4	0.18	**1.3**
Fe_2O_3	0.73	0.38–0.54	2.5–2.8	0–0.95	0.03–0.73	0.21	0.06	0.19	0.3–2	-	0.16–0.4	0.43	**1.5**
CaO	1.4	0.25–1.0	1.0–1.3	1.36–6	0.3–1.4	0.41	0.58	1.4	0.5–2.5	0.06–1.2	0.4–0.49	1.03	**2.2**
K_2O	2.46	0.58–2.0	2.4–2.5	0–0.1	0.7–2.4	2.31	2.54	3.84	0.1–2.3	0.6–2.5	2.0–3.0	0.72	**4.5**
M_gO	0.57	0.12–2.0	0.18–0.28	1.3–1.81	0.1–0.5	1.59	0.26	0.37	0.1–1.8	0.17–0.26	0.35–0.50	0.35	**2.1**
Na_2O	0.11	0–0.15	0.03–0.8	-	0.11–0.2	-	0.09	1.15	0.1–0.5	0–0.3	0.10–1.12	0.02	**0.4**
MnO_2	-	-	-	-	-	-	-	-	-	-	-	-	**0.3**
CuO	-	-	-	-	-	-	-	-	-	-	-	-	**0.01**

[1] [37], [2] This work.

The results of measurements with metal ion mixtures show that Cu has a highly competitive behavior and is retained to a greater extent than the other metals. Particle size also plays a significant role. Cu was quantitatively fixed with the particle size fraction between 90 and 300 µm whereas the adsorption is 98% and 80% with the <90 µm and >300 µm fraction, respectively. The adsorption of Cd and Mn in mixtures decreases compared to the values obtained with the individual metals, due to the intense competition of Cu. The highest Cd adsorption is in the fraction between 90 and 300 µm, and Mn adsorption in the fraction <300 µm and in the fraction between 90 and 300 µm is slightly lower.

From these results, it can be determined that the adsorption depends not only on particle size. The fraction >300 µm, having the lowest surface area, shows, as expected, the lowest absorption capacity. Nevertheless, low retention was observed also for the smallest particle size fraction (<90 µm), probably because of the difficulty in percolating the solution through the particle bed. Therefore, a dependence of retention on particle size can be observed for >90 µm for Cd and Mn, probably due to their lower affinity to cation exchange sites on the particle surface. Basing on these results, the particle size fraction between 90 and 300 µm was used for the subsequent tests.

Contaminant uptake will be increased with a larger surface area or smaller size of rice husk, as already reported in Shamsollahi et al. (2019) [38]. Many studies on rice husk as a biosorbent on heavy metals in the literature confirm its use in the particle size range found. A range between 250 and 297 µm was selected by Ajmal et al. (2003), 300 µm by Bishnoi et al. (2004), and 150 µm by Srivastava et al. (2006) [39–41].

3.2.2. Effect of pH and Buffer Concentration

Figure 4 shows the results of the effect of pH and buffer concentration on the retention of Cd, Cu, and Mn.

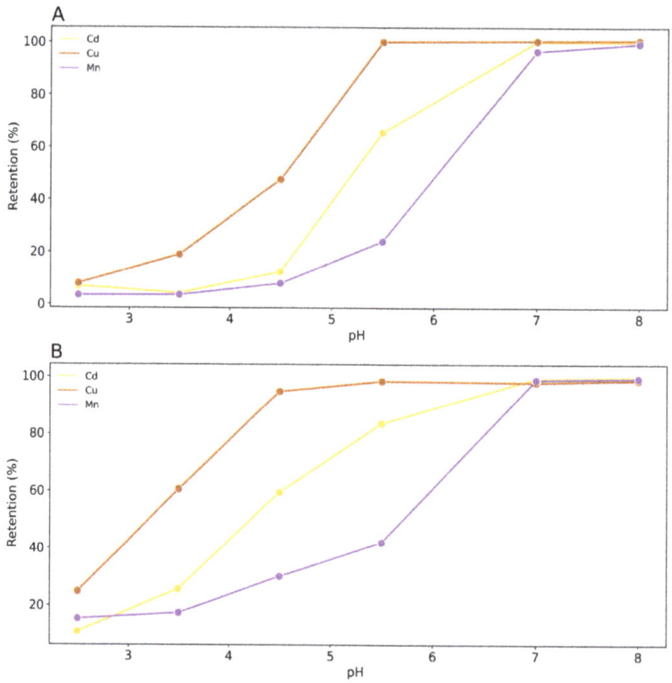

Figure 4. Retention of Cd, Cu, and Mn on rice husk as a function of pH (initial concentration of metals 1×10^{-4} M): (**A**) buffer concentration 1×10^{-1} M; (**B**) buffer concentration of 1×10^{-2} M.

As expected, the retention of all metals decreased with the pH due to the progressive protonation of the cationic exchange sites present on the rice husk surface that are less available to retain the metals. Moreover, the retention decreases with the increasing of the buffer concentration, linked to the concentration of Na^+, owing to the competition with other metals for adsorption sites. It is worth noting that in both cases of different ionic strengths, pH has a strong influence on metal retention. This means that the mechanism of interaction between rice husk and the metals could involve a formation of outer-sphere complexes that are pH-dependent [42]. It can be concluded that low buffer concentration is optimal for the retention of metals. Considering the different behavior of metals, Cu exhibits a retention of 99% at pH 5.5, regardless of the concentration of the buffer, followed by Cd, 64% and 32%, and Mn, 18% and 12%, for a buffer concentration of 1×10^{-2} M and 1×10^{-1} M, respectively. If we consider the hydroxide complex formation constant of the metals investigated in terms of pK_f, we have Cu = 7.9; Cd = 10.1; and Mn = 10.6. This result may be indicative of a specific adsorption mechanism involving the exchange of metal cations with the surface ligands of the rice husk, which can partially form covalent bonds with these ions. In fact, the metals most able to form hydroxy complexes are specifically absorbed to the greatest extent [43]. But this is an explainable reason only for pH 5–8, where the hydroxy complexes are favored. However, the affinity series identified in the pH range of 5.5–8.0 does not appear to be related to the solubility of metal hydroxides. Indeed, according to the values of the solubility product constants (pK_{sp}), $Cu(OH)_2$ = 19.66; $Cd(OH)_2$ = 13.60; and $Mn(OH)_2$ = 12.72; pK_{sp} is only exceeded at pH 7.0 and pH 8.0 for $Cu(OH)_2$ [44]. However, even at these pH levels, no precipitate was observed.

Considering only the interaction of metal aqueous species and silanol groups, the possibility of a mere electrostatic interaction can be ruled out, as the three metal cations are presented with the same charge. Meanwhile, if we consider the covalent index $X_m{}^2r$, where X_m is the electron-attracting capability of an atom in a molecule based on Pauling's electronegativity and r is the ionic radius, the values for the three metals studied are in the following order: $Cu^{2+} > Cd^{2+} > Mn^{2+}$. These results could again justify the order of retention exhibited in the experiments where Mn is the least retained followed by Cd and Cu.

3.2.3. Effect of the Presence of Ligands

The behavior of heavy metals in the presence of organic ligands toward their adsorption on rice husk is crucial since organic compounds are typically present in natural waters, industrial and processing waters, soils, sediments, and wastewaters, in addition to heavy metals [45].

First, we chose a group of carboxylic acids with varying behaviors and increasing complexity: oxalic, malonic, succinic, glutaric, tartaric, and citric acids. We also took into account ligands that can be added to natural systems by anthropogenic activities, such as EDTA and NTA. For instance, the extensive use of EDTA as a chelate in industrial and agricultural settings raises the concentration of the chemical in a variety of water sources and wastewaters that could affect soils with accidental floods and spills [46].

Given the complexity of the system, the distribution of species in the solution under the experimental conditions used was calculated using PyEs software (Table S4). The main forms of the ligands at the pH studied are as follows: H_2EDTA^{2-}, $HNTA^{2-}$, $HTart^-$, $HOxa^-$, $HMal^-$, H_2Succ, H_2Glut and H_2Citr^-. For completeness, Table S5 summarizes the pKa of the ligands and the formation constant of the complexes between the metals and the ligands.

Figure 5 shows the results obtained in the presence of the different ligands compared with the experiment led with the same conditions in the absence of ligands.

ANOVA results in Table S6 reported that for all metals, there was a statistically significant difference between groups. The presence of ligands with high complexation constants, such as NTA and EDTA, significantly prevents Cu from being retained by rice husk. On the other hand, the presence of NTA increases Mn retention by rice husk.

Although the formation constant of the Mn-NTA complex has a higher value than all the complexes with other ligands (excluding EDTA), it is the lowest compared to Cd-NTA and Cu-NTA. If we consider the results of PyEs in Table S4, Mn in the presence of NTA is mainly free in the form of ions and there is no formation of the metal–ligand complex. This leads to the greater retention of Mn on the rice husk in the presence of that ligand.

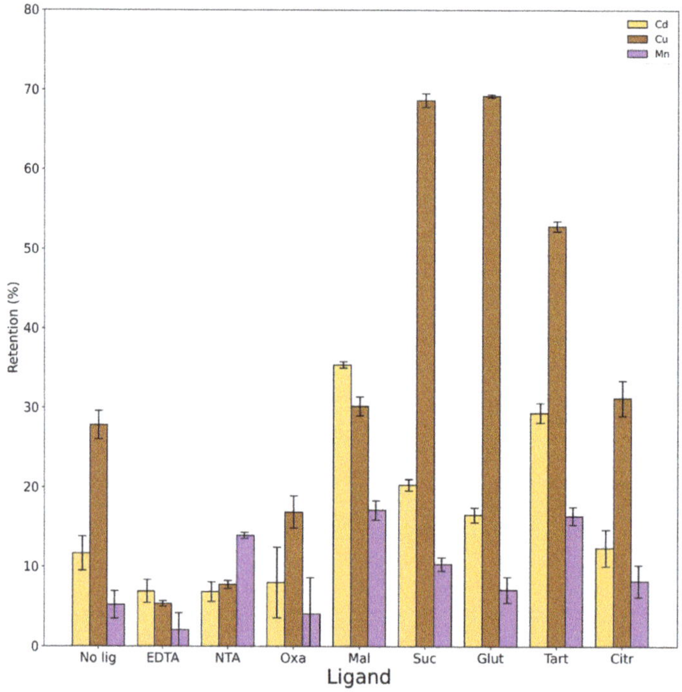

Figure 5. Percentages of retention of Cd, Cu, and Mn on rice husk obtained in the experiments with the different ligands compared to with the experiment with no ligand. Black bars represent confidence interval (95%).

Considering four dicarboxylic acids (oxalic acid, malonic acid, succinic acid, and glutaric acid), there is a slight correlation between the results and the length of their aliphatic chain. Ligands with smaller carbon chains (oxalic and malonic acids) have formation constants with all three metals higher than those of ligands with longer structures (succinic and glutaric acids). Therefore, the formation of metal complexes between smaller carboxylic acids and the metals should be favored and should lead to a decrease in the retention of the metal itself on the rice husk. In fact, if we report only statistically significant differences (p value < 0.01), Cu exhibits a retention value of 27% in the absence of ligands, while exhibiting values of 68 and 69% in the presence of succinic and glutaric acids, and a value of 17% in the presence of oxalic acid. This is confirmed by the highest percentage of free Cu in the presence of succinic and glutaric acid, also due to the fact that at a pH of 3.5, these ligands are not dissociated, whereas a formation of complexes like ML and ML_2 with oxalic acid is observed that prevents the retention of Cu by rice husk.

Considering Mn and Cd with the four dicarboxylic acids, the formation of complexes between the metal and the ligands is not favored. In fact, both metals exhibit a slight increase in the retention on rice husk in the presence of ligands that is statistically significant (p value < 0.01) for Mn with malonic acid (from 5% to 17%) and for Cd with malonic acid (from 12% to 35%) and succinic acid (from 12% to 20%).

Although it appears that tartaric acid does not form complexes with any of the three metals investigated, experimental results show a significant increase in the retention with the presence of this ligand (from 5% to 17% for Mn, from 12% to 30% for Cd and from 27% to 52% for Cu). It cannot be ruled out that there may be an interaction between the rice husk exchange sites with the ligand in question that may facilitate greater metal retention in respect of the results given by the distribution of the species calculated by PyEs. In the study of Wong et al. (2003), in which different modifications of rice husk by various carboxilic acids (namely oxalic, malic, citric, and tartaric acids) were tested to improve metal sorptions from aqueous solutions, the results showed that the best were those obtained with rice husk modified with tartaric acid, shedding light on how this ligand interacts particularly with this matrix [47].

Finally, citric acid seems to not have an influence on the retention of the metals.

In summary, the results showed that the nature of the ligands affects the metal retention percentage on the rice husk. The effect of the ligands is different and depends on the structure of the ligands themselves but also on the type of metal, as well as on the pH of the system. Cu turns out to be the most variable among the three metals and the one that forms more stable complexes with all the ligands.

3.2.4. Total Retention Capacity

After 150 mL of elution, the rice husk reached saturation with all the metals investigated. The breakthrough values show that the total retention capacity decreases in the following order: Cd (12.00 mg/g or 21.4 meq/100 g) > Cu (6.60 mg/g or 21.0 meq/100 g) > Mn (4.00 mg/g or 14.6 meq/100 g). This order is confirmed by a study of Krishnani et al. that investigated the capacity of alkali-treated rice husk on different metal ions adsorption [48]. They explained that lignin and cellulose, as major components of rice husk, have functional groups, i.e., alcohol, carboxyl, and ketone which interact with metallic cations. They found values of adsorption on rice husk of 14.4, 10.8, and 7.7 mg/g for Cd, Cu and Mn, respectively, for an initial concentration of the metal of 2.0×10^{-4} M at pH = 5.5. Other studies have focused on the conversion of rice husk into biochar for using it as an adsorbent for organic and inorganic pollutants. Bao (2023), Pellera et al. (2012), and Xu et al. (2013) found a value of 2.92 and 4.96 mg/g for Cu adsorption on biochar obtained from rice husk in the pH range 5–6 and an initial concentration of the metal of 0.1 M and 5.0×10^{-3} M [36,49,50].

Considering other types of materials that can be used as adsorbents, summarized in Table 2, it can be seen that eggshell wastes exhibit a higher capacity for the absorption of Cd and Cu than rice husk. On the other hand, the total retention capacity of rice husk towards Cd is greater than that of Na-montmorillonite and comparable to that of wheat straw and coal fly ash. The total retention capacity of rice husk for Cu is greater than that of Na–montmorillonite and wheat shell, but smaller than that of the other sorbents. In general, rice husk was found to have a good total capacity toward all considered metals.

Table 2. Total capacity calculated for Cd, and Cu expressed in mg/g for some materials.

	Total Capacity (mg/g)		**References**
	Cd	**Cu**	
Banana peel	30.7	49.5	[51]
Cashew nutshell	22.11	-	[52]
Wheat shell	-	0.83	[53]
Wheat straw	14.56	11.43	[54]
Eggshell waste	111.1	142.6	[55]
Coal fly ash	19.98	20.92	[56]
Na–montmorillonite	5.20	3.04	[57]
Rice husk	12	6.6	This work

3.3. Implementation in Semi-Field Conditions

3.3.1. Effect on Mobility and Reactivity of Heavy Metals in Soil

To investigate the impact of rice husk on the mobility and reactivity of heavy metals in soil, we utilized a contaminated soil previously characterized by Malandrino et al. (2011) [28]. The contaminated soil exhibited notable contamination with heavy metals, surpassing regulatory thresholds established by "DECRETO 1 marzo 2019, n. 46" for agricultural soils in Italy [58]. Cr concentrations were found to be 3262 mg/kg, greatly exceeding the threshold limit of 150 mg/kg. Cu was 2777 mg/kg, surpassing the threshold of 200 mg/kg, while Ni levels reached 624 mg/kg, exceeding the limit of 120 mg/kg. Pb was also present at a concentration of 687 mg/kg, (threshold of 100 mg/kg), while the Zn content was 364 mg/kg, just above the threshold of 300 mg/kg.

In this previous work, the distribution of the heavy metals among the fractions obtained by applying Tessier's protocol to Borgomanero soil was investigated. The results revealed higher concentrations of metals in the first two fractions in samples from the center of the site, indicating increased pollution levels. Particularly notable were the elevated percentages of Cd, Cu, Ni, Pb, and Zn in these fractions, attributable to anthropogenic sources. The large $MgCl_2$-extractable fraction of these metals was influenced by the low soil pH. The percentages of metals in the first two fractions were significantly higher in the contaminated soil (Figure S2) compared to the control soil (Figure S3). Additionally, most heavy metals were extracted into the third fraction, with Cu being an exception, primarily found in the fourth fraction due to its complexation with organic matter. Cr behavior suggested a possible input in the environment as Cr(VI), a more mobile and toxic form, which could be reduced to Cr(III) in acidic soils, potentially adsorbed onto Fe oxides. Mn exhibited low extraction percentages in the first four fractions, indicating strong binding to the soil matrix and natural origin. This differentiation underscores the higher mobility of anthropogenic elements compared to lithogenic ones, suggesting potential environmental release under changing soil conditions [43]. The acidic environment found in this soil can significantly affect metal mobility. In fact, as pH decreases, the solubility of cationic species increases and H^+ ions can compete with cations for binding sites in soil particles. Another critical factor is the exceptionally high organic matter content (30% w/w) of the contaminated soil. Repeated flooding from a nearby stream probably caused this accumulation. The original creek, now diverted, received industrial effluent, including that from electroplating plants. The high organic matter content can have a complex effect on metal mobility. On the one hand, organic matter can function as a ligand, immobilizing metals and reducing their availability. On the other hand, the decomposition of organic matter can also release complexing agents that can increase metal mobility improving their solubility also at higher pH values. Furthermore, the soil has a high cation exchange capacity (CEC) of 31.3 cmol/kg. CEC represents the ability of the soil to hold positively charged ions, including metals. A high CEC generally means a greater capacity to retain metals, potentially reducing their mobility and subsequent environmental risks.

As shown in Figure 6, the addition of rice husk as a soil amendment resulted in a noticeable change in the distribution of heavy metals. A clear decrease in the concentration of Ni, Zn, and Pb was observed in the mobile fraction, suggesting that they may have been immobilized by rice husk. In addition to the high sorption capacity, rice husk can increase soil surface and porosity, further favoring metal retention and decreasing bioavailability to plant species. In addition, the introduction of rice husk displaced Cr and Cu into the strongly bound fractions compared to the unamended contaminated soil, likely due to chelation mechanisms with rice husk components. Finally, rice husk increased soil Mn mobility, probably due to the presence of Mn in the rice husk or to soil variability.

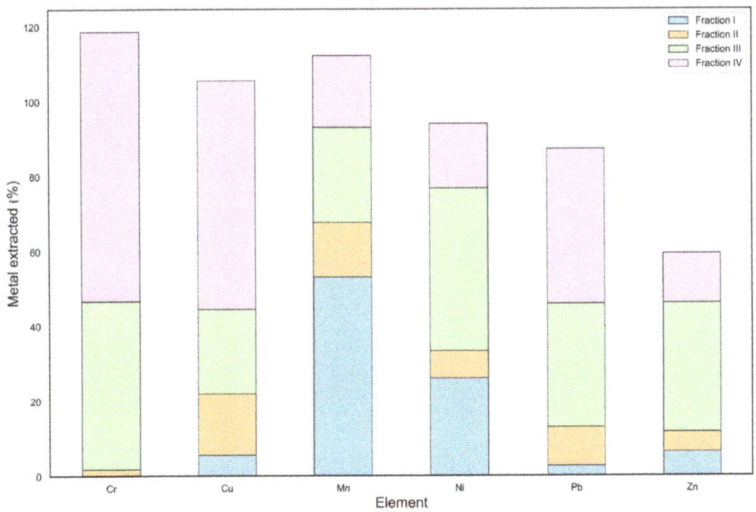

Figure 6. Heavy metal percentages extracted into the first four fractions according to Tessier's protocol for contaminated soil amended by rice husk.

The Tessier sequential extraction was used to evaluate the availability of metals within the Borgomanero soil and to predict the uptake propensity of certain metals by plants. Significant positive correlations were found between the cumulative concentrations of Ni, Zn, and Pb in the initial two Tessier fractions and their corresponding concentrations in plant tissues (Ni − R = 0.83, $p < 0.01$; Zn − R = 0.97, $p < 0.01$; Pb − R = 0.87, $p < 0.01$). In contrast, Cu showed a weak positive correlation (R = 0.17, $p > 0.05$), suggesting that the fraction available to plants was not related to the first two Tessier fractions. Interestingly, there appeared a negative correlation for Mn (R = −0.43, $p > 0.05$) and no significant correlation was found for Cr (Cr − R = −0.09, $p > 0.05$), suggesting that the first two Tessier fractions may not accurately reflect plant-accessible Mn and Cr species. It is worth remembering that both metals can be present in soil at different oxidation states, influencing their mobility and bioavailability. It is possible that the first two Tessier fractions allow for a less accurate estimate of the pool available to plants for those metals that are mostly present in the soil in different ionic forms. These observations underscore the necessity for further research to refine the Tessier protocol or explore alternative methodologies to effectively measure the bioavailability of Cu, Mn, and Cr in soil.

3.3.2. Effect on Lettuce and Spinach Heavy Metal Uptake

As shown in Table 3, lettuce and spinach plants grown in the heavy metal-contaminated soil had significantly higher metal concentrations than the control group grown in the uncontaminated soil. This indicates their ability to absorb and concentrate metals in their tissues under contaminated conditions. To evaluate the effectiveness of rice husk in decreasing the heavy metal uptake by vegetables, we calculated the "reduction efficiency" as follows: ([Me]C − [Me]RH)/[Me]C × 100. The incorporation of rice husk resulted in a significant reduction in heavy metal uptake by plants. In fact, by comparing the plants grown in the treated soils with the ones grown in the untreated contaminated soil, the reduction in heavy metal uptake ranged from 40 to 60% for Mn and Zn to almost complete elimination (close to 100%) for Cr, Cu, Ni, Cd, and Pb. These results are particularly noteworthy for Cr, Cu, and Ni, since their concentrations in the contaminated soil significantly exceeded the Italian limits.

Table 3. Concentrations of heavy metals (mg/kg dry weight) in plants, including data for lettuce and spinach cultivated in the control soil, in contaminated soil, and in contaminated soil amended with rice husk.

Plant	Metal	Control			Contaminated Soil			Contaminated Soil + Rice Husk			Reduction Efficiency
Lettuce	Cd		<0.004		2.85	±	0.30	0.38	±	0.02	87%
	Cr	0.18	±	0.047	3094	±	40	5.93	±	0.42	100%
	Cu	1.94	±	1.334	3254	±	215	18.6	±	15.2	99%
	Mn	6.14	±	0.393	289.2	±	7.5	171	±	17	41%
	Ni	0.97	±	0.43	606.5	±	25.5	68.7	±	0.64	89%
	Pb		<0.125		632.9	±	35.9	2.81	±	2.05	100%
	Zn	26.98	±	1.064	352.1	±	0.2	198	±	28.7	44%
Spinach	Cd		<0.004		2.57	±	0.27	0.97	±	0.03	74%
	Cr	1.327	±	0.022	3011	±	77	14.6	±	3.62	100%
	Cu	3.717	±	0.457	3062	±	6	109	±	3.39	98%
	Mn	5.844	±	0.154	304.9	±	8.3	68.3	±	7.23	59%
	Ni	1.288	±	0.291	589.1	±	6.7	63.4	±	0.43	89%
	Pb		<0.125		625.7	±	0.7	5.98	±	7.52	99%
	Zn	136.7	±	3.117	347.8	±	0.4	201	±	26.2	43%

Furthermore, the concentration of Pb, Cd, and Zn in the plants were found to be below the critical concentration limit for a 10% yield loss of the plant [43,59]. This is very important because it indicates that the concentration of these heavy metals does not result in a loss of crop yield. On the other hand, Ni, Cr, and Cu remained at levels of concern; however, the significant reduction observed compared to the initial highly contaminated conditions is a promising result. Specifically, the concentration of Cu in lettuce plants grown in soil amended with rice husk was similar to the baseline values found in the literature, demonstrating that rice husk may be an effective amendment for reducing Cu uptake by lettuce, confirming the high capability of the rice husk to retain Cu reported in the above sections [59].

The assessment of the plant capacity to absorb chemical elements from its growth medium, as determined by the Bioaccumulation Factor (BAF), showed that for Cd, BAF values exceeded 10, indicative of a pronounced propensity for accumulation within the plant, owing to the prevalence of mobile chemical forms that are readily accessible to plant uptake. In contrast, Zn, Cu, and Pb typically exhibit BAF values ranging between 1 and 0.1, signifying an intermediate level of accumulation within the plant tissue, attributable to the presence of chemical forms that are only moderately mobile. Meanwhile, Mn, Cr, and Ni typically display BAF values between 0.1 and 0.01, indicating a comparatively modest degree of accumulation within the plant, owing to the tendency of these metals to occur in forms that are relatively immobile and less accessible to plants (Cr and Ni) or present in large quantities in soils (Mn) [59].

As shown in Figure 7, BAF values were consistently below for plants grown in uncontaminated soils. This is likely due to the elevated organic matter content and substantial cation exchange capacity inherent to the control soil. Only Zn accumulated in spinach, which suggests an intrinsic predisposition of this plant toward the assimilation and accumulation of Zn. Plants cultivated in contaminated soil matrices typically exhibit BAF values approaching one, underscoring the accumulator behavior of lettuce and spinach in response to heightened concentrations and mobility of heavy metals within the soil environment.

The incorporation of rice husk as an amendment significantly reduces BAF values for most heavy metals. It is noteworthy that BAF values for Zn, Cr, and Ni return to baseline levels, due to the high immobilization determined by rice husk. Furthermore, BAF values for Cu, Pb, and Cd are significantly lower than the normal levels of bioaccumulation, indicating the effectiveness of chemical immobilization mechanisms in reducing the accumulation of these heavy metals in plant tissues [59]. Finally, the reduction in BAF values for Mn with the incorporation of rice husk is lower than those of other metals: this outcome may be explained by the lower affinity of Mn for rice husk, as explained in the previous sections of this study.

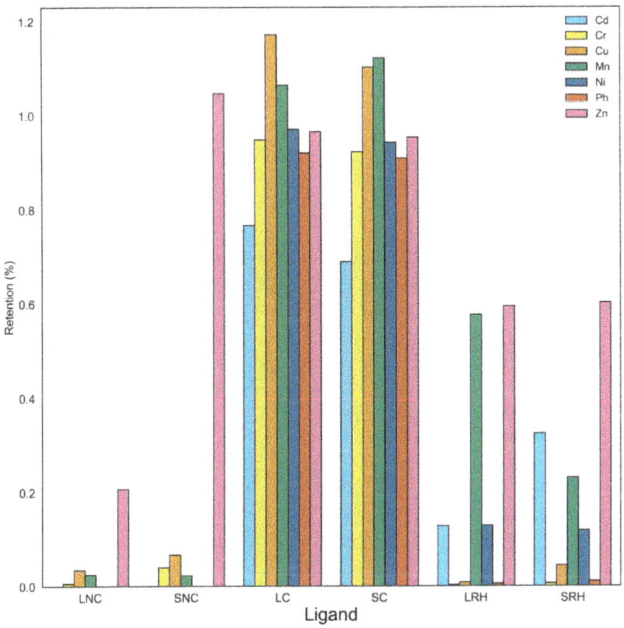

Figure 7. Bioaccumulation factors (BAFs) of heavy metals by plants, including data for lettuce and spinach cultivated under various conditions: in uncontaminated soil (LNC for lettuce, SNC for spinach), in contaminated soil (LC for lettuce, SC for spinach), and in contaminated soil amended with rice husk (LRH for lettuce, SRH for spinach).

4. Conclusions

Rice husk has proven to be an efficient matrix in removing metal ions from aqueous solutions. Its retention depends on solution pH, buffer concentration, and the particle size of the rice husk. The highest retention efficiency was achieved with an intermediate rice husk particle size (90–300 μm), 1.0×10^{-2} M buffer concentration, and pH > 5.5. Organic acids that act as ligands may control the interaction of metals with the rice husk, depending on their nature and complex constants. In particular, this study quantitatively demonstrated that rice husk exhibits retention capacities of up to 100% for Cu, 64% for Cd, and 18% for Mn. This makes rice husk a promising and sustainable material for the amendment of contaminated soils, due also to its availability in massive quantities and at low cost.

The decrease in plant heavy metal uptake observed after rice husk application highlights its potential as a viable solution for mitigating metal pollution and safeguarding food safety. The pot experiments showed reductions in heavy metal uptake by *Lactuca sativa* and *Spinacia oleracea* of 40–60% for Mn and Zn, and nearly complete immobilization (up to 100%) of Cd, Cu, and Pb. Additionally, the modification of heavy metal distribution within soil fractions after amendment addition emphasizes the transformative role of rice husk in immobilizing metals and reducing their bioavailability. When compared with other common remediation media such as activated carbon or chemically treated biochar, rice husk offers a simpler, cost-effective alternative that is especially suitable for areas with significant agricultural production of rice, where this byproduct is abundant. However, it is important to note that while the rice husk was highly effective in immobilizing Cd, Cu, and Pb, its efficiency in reducing Mn mobility was lower, indicating a potential limitation for soils heavily contaminated with Mn. Additionally, although the pH-dependent nature of metal retention suggests broad applicability in soils with acidic to neutral pH, its performance in alkaline soils or soils with high organic matter content requires further investigation. However, further research is needed to refine remediation strategies and accurately predict metal bioavailability. Future studies should focus on long-term field trials to assess the durability of rice husk's immobilization capacity under varying environmental conditions, such as rainfall, temperature fluctuations, and soil microbial activity. In conclusion, the chemical stabilization by rice husk evidenced in this study demonstrated to be a simple and cost-effective remediation technique that allows for the reduction in heavy metal assimilation from contaminated soils by edible plants and holds promise for sustainable agricultural practices applicable in urban and peri-urban soils.

Supplementary Materials: The following supporting information can be downloaded at: https://www.mdpi.com/article/10.3390/toxics12110790/s1, Table S1: Wavelengths chosen of the analytes determined and the limit of detections (*LOD*, µg/L); Table S2: Characteristics of the contaminated soil of Borgomanero and control soil; Table S3: Concentrations with standard deviation of the analytes determined in the rice husk, expressed in mg/kg; Table S4: Percentage of the formation of the species with the different ligands for Cu, Cd, and Mn, calculated by the software PyES. For each experiment the concentration of the metal is 1.0×10^{-4} M, the concentration of the ligand 3.0×10^{-4} M, and the concentration of the buffer acetate 1.0×10^{-2} M; Table S5: Equilibrium constants of the considered ligands [60–62]; Table S6: ANOVA table for each metal and its significance between the groups; Figure S1: A map of the location of the soils contaminated: on the top, the position of the site in the European continent; on the bottom, the position of the site in the Italian region; Figure S2: Heavy metal percentages extracted into the first four fractions according to Tessier's protocol for contaminated soil; Figure S3: Heavy metal percentages extracted into the first four fractions according to Tessier's protocol for non-contaminated soil.

Author Contributions: Writing—original draft: R.C. and A.D.; writing—review and editing: R.C., A.D., A.G., O.A., P.I., L.F., S.B., S.C., L.C. and M.M.; conceptualization: M.M.; methodology: M.M. and L.C.; validation: M.M., L.C. and S.B.; formal Analysis: R.C. and A.D.; investigation: R.C. and A.D.; resources: M.M., L.C., A.G. and O.A.; funding acquisition: M.M. and L.C.; supervision: M.M. and L.C.; project administration: M.M. All authors have read and agreed to the published version of the manuscript.

Funding: This study was carried out within the Agritech National Research Center and received funding from the European Union Next-GenerationEU (PIANO NAZIONALE DI RIPRESA E RESILIENZA (PNRR)—MISSIONE 4 COMPONENTE 2, INVESTIMENTO 1.4—D.D. 1032 17/06/2022, CN00000022). This manuscript reflects only the authors' views and opinions; neither the European Union nor the European Commission can be considered responsible for them.

Data Availability Statement: Data will be made available on request.

Acknowledgments: The authors want to thank Simone Centoz and Thangaraja Chinnathangavel for their help in the sample preparation and analysis.

Conflicts of Interest: The authors declare no conflicts of interest. The founders had no role in the design of the study; in the collection; analysis or interpretation of data; in the writing of the manuscript; or in the decision to publish the results.

References

1. Kiran, B.R.; Prasad, M.N.V. Rice husk and wood derived charcoal for remediation of metal contaminated soil. In *Handbook of Assisted and Amendment: Enhanced Sustainable Remediation Technology*; John Wiley & Sons: Hoboken, NJ, USA, 2021; pp. 235–266, ISBN 978-1-119-67039-1.
2. Laidlaw, M.A.S.; Filippelli, G.M.; Brown, S.; Paz-Ferreiro, J.; Reichman, S.M.; Netherway, P.; Truskewycz, A.; Ball, A.S.; Mielke, H.W. Case Studies and Evidence-Based Approaches to Addressing Urban Soil Lead Contamination. *Appl. Geochem.* **2017**, *83*, 14–30. [CrossRef]
3. Pavel, L.V.; Gavrilescu, M. Overview of Ex Situ Decontamination Techniques for Soil Cleanup. *Environ. Eng. Manag. J.* **2008**, *7*, 815–834. [CrossRef]

4. Gupta, U.C.; Gupta, S.C. Trace Element Toxicity Relationships to Crop Production and Livestock and Human Health: Implications for Management. *Commun. Soil Sci. Plant Anal.* **1998**, *29*, 1491–1522. [CrossRef]
5. Sarker, A.; Masud, M.A.A.; Deepo, D.M.; Das, K.; Nandi, R.; Ansary, M.W.R.; Islam, A.R.M.T.; Islam, T. Biological and Green Remediation of Heavy Metal Contaminated Water and Soils: A State-of-the-Art Review. *Chemosphere* **2023**, *332*, 138861. [CrossRef]
6. Shahid, M.; Dumat, C.; Khalid, S.; Schreck, E.; Xiong, T.; Niazi, N.K. Foliar Heavy Metal Uptake, Toxicity and Detoxification in Plants: A Comparison of Foliar and Root Metal Uptake. *J. Hazard. Mater.* **2017**, *325*, 36–58. [CrossRef]
7. Gurdon, L.; Esmahi, L.; Amponsah, N.Y.; Wang, J. Life Cycle Cost Analysis of Contaminated Site Remediation Using Information Technology Tools. *Environ. Dev. Sustain.* **2021**, *23*, 10173–10193. [CrossRef]
8. Cai, M.; McBride, M.B.; Li, K.; Li, Z. Bioaccessibility of As and Pb in Orchard and Urban Soils Amended with Phosphate, Fe Oxide and Organic Matter. *Chemosphere* **2017**, *173*, 153–159. [CrossRef]
9. El Rasafi, T.; Haouas, A.; Tallou, A.; Chakouri, M.; Aallam, Y.; El Moukhtari, A.; Hamamouch, N.; Hamdali, H.; Oukarroum, A.; Farissi, M.; et al. Recent Progress on Emerging Technologies for Trace Elements-Contaminated Soil Remediation. *Chemosphere* **2023**, *341*, 140121. [CrossRef]
10. Komárek, M.; Vaněk, A.; Ettler, V. Chemical Stabilization of Metals and Arsenic in Contaminated Soils Using Oxides—A Review. *Environ. Pollut.* **2013**, *172*, 9–22. [CrossRef]
11. Kumpiene, J.; Lagerkvist, A.; Maurice, C. Stabilization of As, Cr, Cu, Pb and Zn in Soil Using Amendments—A Review. *Waste Manag.* **2008**, *28*, 215–225. [CrossRef]
12. Bolan, N.; Kunhikrishnan, A.; Thangarajan, R.; Kumpiene, J.; Park, J.; Makino, T.; Kirkham, M.B.; Scheckel, K. Remediation of Heavy Metal(Loid)s Contaminated Soils—To Mobilize or to Immobilize? *J. Hazard. Mater.* **2014**, *266*, 141–166. [CrossRef] [PubMed]
13. Luh, B.S. *Rice, Volume 2: Utilization*; Springer: New York, NY, USA, 1991; ISBN 978-0-442-00485-9.
14. Karić, N.; Maia, A.S.; Teodorović, A.; Atanasova, N.; Langergraber, G.; Crini, G.; Ribeiro, A.R.L.; Dolić, M. Bio-Waste Valorisation: Agricultural Wastes as Biosorbents for Removal of (in)Organic Pollutants in Wastewater Treatment. *Chem. Eng. J. Adv.* **2022**, *9*, 100239. [CrossRef]
15. Li, Z.; Zheng, Z.; Li, H.; Xu, D.; Li, X.; Xiang, L.; Tu, S. Review on Rice Husk Biochar as an Adsorbent for Soil and Water Remediation. *Plants* **2023**, *12*, 1524. [CrossRef]
16. Nayyar, D.; Shaikh, M.A.N.; Nawaz, T. Remediation of Emerging Contaminants by Naturally Derived Adsorbents. In *New Trends in Emerging Environmental Contaminants*; Singh, S.P., Agarwal, A.K., Gupta, T., Maliyekkal, S.M., Eds.; Springer: Singapore, 2022; pp. 225–260. ISBN 9789811683671.
17. Zheng, R.-L.; Cai, C.; Liang, J.-H.; Huang, Q.; Chen, Z.; Huang, Y.-Z.; Arp, H.P.H.; Sun, G.-X. The Effects of Biochars from Rice Residue on the Formation of Iron Plaque and the Accumulation of Cd, Zn, Pb, As in Rice (*Oryza sativa* L.) Seedlings. *Chemosphere* **2012**, *89*, 856–862. [CrossRef]
18. Derakhshan Nejad, Z.; Jung, M.C. The Effects of Biochar and Inorganic Amendments on Soil Remediation in the Presence of Hyperaccumulator Plant. *Int. J. Energy Environ. Eng.* **2017**, *8*, 317–329. [CrossRef]
19. Bian, R.; Shi, W.; Luo, J.; Li, W.; Wang, Y.; Joseph, S.; Gould, H.; Zheng, J.; Zhang, X.; Liu, X.; et al. Copyrolysis of Food Waste and Rice Husk to Biochar to Create a Sustainable Resource for Soil Amendment: A Pilot-Scale Case Study in Jinhua, China. *J. Clean. Prod.* **2022**, *347*, 131269. [CrossRef]
20. Buck, R.P.; Rondinini, S.; Covington, A.K.; Baucke, F.G.K.; Brett, C.M.A.; Camoes, M.F.; Milton, M.J.T.; Mussini, T.; Naumann, R.; Pratt, K.W.; et al. Measurement of pH. Definition, Standards, and Procedures (IUPAC Recommendations 2002). *Pure Appl. Chem.* **2002**, *74*, 2169–2200. [CrossRef]
21. Barquilha, C.E.R.; Braga, M.C.B. Adsorption of Organic and Inorganic Pollutants onto Biochars: Challenges, Operating Conditions, and Mechanisms. *Bioresour. Technol. Rep.* **2021**, *15*, 100728. [CrossRef]
22. De Franco, M.A.E.; de Carvalho, C.B.; Bonetto, M.M.; Soares, R.d.P.; Féris, L.A. Removal of Amoxicillin from Water by Adsorption onto Activated Carbon in Batch Process and Fixed Bed Column: Kinetics, Isotherms, Experimental Design and Breakthrough Curves Modelling. *J. Clean. Prod.* **2017**, *161*, 947–956. [CrossRef]
23. Rudnicki, P.; Hubicki, Z.; Kołodyńska, D. Evaluation of Heavy Metal Ions Removal from Acidic Waste Water Streams. *Chem. Eng. J.* **2014**, *252*, 362–373. [CrossRef]
24. Ross, A.; Willson, V.L. One-Way Anova. In *Basic and Advanced Statistical Tests: Writing Results Sections and Creating Tables and Figures*; Ross, A., Willson, V.L., Eds.; Sense Publishers: Rotterdam, The Netherlands, 2017; pp. 21–24, ISBN 978-94-6351-086-8.
25. Dunnett, C.W. A Multiple Comparison Procedure for Comparing Several Treatments with a Control. *J. Am. Stat. Assoc.* **1955**, *50*, 1096–1121. [CrossRef]
26. Castellino, L.; Alladio, E.; Bertinetti, S.; Lando, G.; De Stefano, C.; Blasco, S.; García-España, E.; Gama, S.; Berto, S.; Milea, D. PyES—An Open-Source Software for the Computation of Solution and Precipitation Equilibria. *Chemom. Intell. Lab. Syst.* **2023**, *239*, 104860. [CrossRef]
27. Böhmer, R.G. Separation and Preconcentration Methods in Inorganic Trace Analysis. J. Minczewski, J. Chwastowska, and R. Dybczyński. Ellis Horwood Ltd., Chichester, Publishers, Distributed by John Wiley & Sons, New York, Chichester, Brisbane, Toronto. 1982. ISBN 0-8531 2-1 65-6 or 0-470-271 69-8, Xi + 543 Pages. *J. High Resolut. Chromatogr.* **1983**, *6*, 208. [CrossRef]

28. Malandrino, M.; Abollino, O.; Buoso, S.; Giacomino, A.; La Gioia, C.; Mentasti, E. Accumulation of Heavy Metals from Contaminated Soil to Plants and Evaluation of Soil Remediation by Vermiculite. *Chemosphere* **2011**, *82*, 169–178. [CrossRef] [PubMed]
29. USD. A Natural Resources Conservation Service, U.S. Department of Agriculture. Available online: https://www.nrcs.usda.gov/ (accessed on 9 July 2024).
30. Tessier, A.; Campbell, P.G.C.; Bisson, M. Sequential Extraction Procedure for the Speciation of Particulate Trace Metals. *Anal. Chem.* **1979**, *51*, 844–851. [CrossRef]
31. Abollino, O.; Giacomino, A.; Malandrino, M.; Mentasti, E.; Aceto, M.; Barberis, R. Assessment of Metal Availability in a Contaminated Soil by Sequential Extraction. *Water Air. Soil Pollut.* **2006**, *173*, 315–338. [CrossRef]
32. Davidson, C.M.; Duncan, A.L.; Littlejohn, D.; Ure, A.M.; Garden, L.M. A Critical Evaluation of the Three-Stage BCR Sequential Extraction Procedure to Assess the Potential Mobility and Toxicity of Heavy Metals in Industrially-Contaminated Land. *Anal. Chim. Acta* **1998**, *363*, 45–55. [CrossRef]
33. Medyńska-Juraszek, A.; Marcinkowska, K.; Gruszka, D.; Kluczek, K. The Effects of Rabbit-Manure-Derived Biochar Co-Application with Compost on the Availability and Heavy Metal Uptake by Green Leafy Vegetables. *Agronomy* **2022**, *12*, 2552. [CrossRef]
34. Ng, K.T.; Herrero, P.; Hatt, B.; Farrelly, M.; McCarthy, D. Biofilters for Urban Agriculture: Metal Uptake of Vegetables Irrigated with Stormwater. *Ecol. Eng.* **2018**, *122*, 177–186. [CrossRef]
35. Soltani, N.; Bahrami, A.; Pech-Canul, M.I.; González, L.A. Review on the Physicochemical Treatments of Rice Husk for Production of Advanced Materials. *Chem. Eng. J.* **2015**, *264*, 899–935. [CrossRef]
36. Bao, J. Chapter 15—Rice. In *ICC Handbook of 21st Century Cereal Science and Technology*; Shewry, P.R., Koksel, H., Taylor, J.R.N., Eds.; Academic Press: Cambridge, MA, USA, 2023; pp. 145–151. ISBN 978-0-323-95295-8.
37. Fapohunda, C.; Akinbile, B.; Shittu, A. Structure and Properties of Mortar and Concrete with Rice Husk Ash as Partial Replacement of Ordinary Portland Cement—A Review. *Int. J. Sustain. Built Environ.* **2017**, *6*, 675–692. [CrossRef]
38. Shamsollahi, Z.; Partovinia, A. Recent Advances on Pollutants Removal by Rice Husk as a Bio-Based Adsorbent: A Critical Review. *J. Environ. Manage.* **2019**, *246*, 314–323. [CrossRef] [PubMed]
39. Ajmal, M.; Ali Khan Rao, R.; Anwar, S.; Ahmad, J.; Ahmad, R. Adsorption Studies on Rice Husk: Removal and Recovery of Cd(II) from Wastewater. *Bioresour. Technol.* **2003**, *86*, 147–149. [CrossRef] [PubMed]
40. Bishnoi, N.R.; Bajaj, M.; Sharma, N.; Gupta, A. Adsorption of Cr(VI) on Activated Rice Husk Carbon and Activated Alumina. *Bioresour. Technol.* **2004**, *91*, 305–307. [CrossRef] [PubMed]
41. Srivastava, V.C.; Mall, I.D.; Mishra, I.M. Characterization of Mesoporous Rice Husk Ash (RHA) and Adsorption Kinetics of Metal Ions from Aqueous Solution onto RHA. *J. Hazard. Mater.* **2006**, *134*, 257–267. [CrossRef]
42. Evangelou, V.P. *Environmental Soil and Water Chemistry*; A Wiley-Interscience Publication: Hoboken, NJ, USA, 2022.
43. Alloway, B.J. *Heavy Metals in Soils: Trace Metals and Metalloids in Soils and Their Bioavailability*; Springer: Dordrecht, The Netherlands, 2012; ISBN 978-94-007-4470-7.
44. Harris, D.C.; Lucy, C.A. *Quantitative Chemical Analysis*, 10th ed.; Macmillan: London, UK, 2010.
45. Malandrino, M.; Abollino, O.; Giacomino, A.; Aceto, M.; Mentasti, E. Adsorption of Heavy Metals on Vermiculite: Influence of pH and Organic Ligands. *J. Colloid Interface Sci.* **2006**, *299*, 537–546. [CrossRef]
46. Darban, A.K.; Foriero, A.; Yong, R.N. Concentration Effects of EDTA and Chloride on the Retention of Trace Metals in Clays. *Eng. Geol.* **2000**, *57*, 81–94. [CrossRef]
47. Wong, K.K.; Lee, C.K.; Low, K.S.; Haron, M.J. Removal of Cu and Pb by Tartaric Acid Modified Rice Husk from Aqueous Solutions. *Chemosphere* **2003**, *50*, 23–28. [CrossRef]
48. Krishnani, K.K.; Meng, X.; Christodoulatos, C.; Boddu, V.M. Biosorption Mechanism of Nine Different Heavy Metals onto Biomatrix from Rice Husk. *J. Hazard. Mater.* **2008**, *153*, 1222–1234. [CrossRef]
49. Pellera, F.-M.; Giannis, A.; Kalderis, D.; Anastasiadou, K.; Stegmann, R.; Wang, J.-Y.; Gidarakos, E. Adsorption of Cu(II) Ions from Aqueous Solutions on Biochars Prepared from Agricultural by-Products. *J. Environ. Manage.* **2012**, *96*, 35–42. [CrossRef]
50. Xu, X.; Cao, X.; Zhao, L. Comparison of Rice Husk- and Dairy Manure-Derived Biochars for Simultaneously Removing Heavy Metals from Aqueous Solutions: Role of Mineral Components in Biochars. *Chemosphere* **2013**, *92*, 955–961. [CrossRef] [PubMed]
51. Li, Y.; Liu, J.; Yuan, Q.; Tang, H.; Yu, F.; Lv, X. A Green Adsorbent Derived from Banana Peel for Highly Effective Removal of Heavy Metal Ions from Water. *RSC Adv.* **2016**, *6*, 45041–45048. [CrossRef]
52. SenthilKumar, P.; Ramalingam, S.; Abhinaya, R.V.; Kirupha, S.D.; Vidhyadevi, T.; Sivanesan, S. Adsorption Equilibrium, Thermodynamics, Kinetics, Mechanism and Process Design of Zinc(II) Ions onto Cashew Nut Shell. *Can. J. Chem. Eng.* **2012**, *90*, 973–982. [CrossRef]
53. Basci, N.; Kocadagistan, E.; Kocadagistan, B. Biosorption of Copper (II) from Aqueous Solutions by Wheat Shell. *Desalination* **2004**, *164*, 135–140. [CrossRef]
54. Dang, V.B.H.; Doan, H.D.; Dang-Vu, T.; Lohi, A. Equilibrium and Kinetics of Biosorption of Cadmium(II) and Copper(II) Ions by Wheat Straw. *Bioresour. Technol.* **2009**, *100*, 211–219. [CrossRef]
55. Zheng, W.; Li, X.; Yang, Q.; Zeng, G.; Shen, X.; Zhang, Y.; Liu, J. Adsorption of Cd(II) and Cu(II) from Aqueous Solution by Carbonate Hydroxylapatite Derived from Eggshell Waste. *J. Hazard. Mater.* **2007**, *147*, 534–539. [CrossRef]

56. Papandreou, A.; Stournaras, C.J.; Panias, D. Copper and Cadmium Adsorption on Pellets Made from Fired Coal Fly Ash. *J. Hazard. Mater.* **2007**, *148*, 538–547. [CrossRef]
57. Abollino, O.; Aceto, M.; Malandrino, M.; Sarzanini, C.; Mentasti, E. Adsorption of Heavy Metals on Na-Montmorillonite. Effect of pH and Organic Substances. *Water Res.* **2003**, *37*, 1619–1627. [CrossRef]
58. DECRETO 1 Marzo 2019, n. 46. Available online: https://www.normattiva.it/uri-res/N2Ls?urn:nir:ministero.ambiente.e.tutela.territorio.e.mare:decreto:2019-03-01;46!vig= (accessed on 9 July 2024).
59. Kabata-Pendias, A. *Trace Elements in Soils and Plants*, 3rd ed.; CRC Press: Boca Raton, FL, USA, 2000; ISBN 978-0-429-19112-1.
60. Martell, A.E.; Smith, R.M. *Critical Stability Constants: First Supplement*; Springer: Boston, MA, USA, 1982; ISBN 978-1-4615-6763-9.
61. Smith, R.M.; Martell, A.E. *Critical Stability Constants*; Springer: Boston, MA, USA, 1989; ISBN 978-1-4615-6766-0.
62. Anderegg, G. Critical Survey of Stability Constants of NTA Complexes. *Pure Appl. Chem.* **1982**, *54*, 2693–2758. [CrossRef]

toxics

MDPI

Article

The Tolerance Differences of Two Industrial Hemp Varieties Under Lead (Pb) Stress

Yanping Xu [1,2,*], Anuwat Kumpeangkeaw [3], Xia An [4], Xuan Chen [1,2], Yuan Zhang [1,2], Pin Lv [1,2], Qingying Zhang [1,2], Rong Guo [1,2], Qingqing Ji [4,5] and Ming Yang [1,2,*]

1 Industrial Crops Research Institute, Yunnan Academy of Agricultural Sciences, Kunming 650205, China; chenxuan9239@163.com (X.C.); 136151701@163.com (Y.Z.); lvpinhemp@163.com (P.L.); hempzhangqingying@126.com (Q.Z.); grmm0207@126.com (R.G.)
2 Yunnan Key Laboratory of Genetic Improvement of Herbal Oil Crops, Kunming 650205, China
3 Department of Agriculture, Ministry of Agriculture and Cooperatives, Bangkok 10900, Thailand; flaaaay66@gmail.com
4 Zhejiang Xiaoshan Institute of Cotton & Bast Fiber Crops Research, Zhejiang Institute of Landscape Plants and Flowers, Zhejiang Academy of Agricultural Science, Hangzhou 311251, China; anxia@zaas.ac.cn (X.A.); jiqingqing@stu.ynu.edu.cn (Q.J.)
5 School of Agriculture, Yunnan University, Kunming 650500, China
* Correspondence: xyp@yass.org.cn (Y.X.); ym@yass.org.cn (M.Y.)

Abstract: Industrial hemp is a crop with a high tolerance and accumulation of lead (Pb). Improving the Pb tolerance and accumulation capacity of industrial hemp is of great scientific and practical importance. This study utilized a pot with soil contaminated with Pb to investigate the differences in Pb tolerance between two industrial hemp varieties, Yunma1 (YM) and Shaanxi Industrial Hemp (SM), under Pb stress. The results indicated that Pb mainly accumulates in the roots of YM and SM (70–80%), with YM having a higher Pb accumulation than SM. It is worth nothing that under high Pb concentration conditions (5000 mg/kg), the Pb accumulation capacity of YM is twice that of SM. Accumulation characteristics of Pb in different plant tissues followed the pattern: roots > stems > leaves > fibers > seeds. In YM, approximately 70% of the absorbed Pb was fixed in the roots and 30% was transported to the above-ground parts. In contrast, SM transported more than 50% of absorbed Pb by roots to the above-ground areas, causing some degree of damage to stems and leaves. Even when Pb concentrations exceed 4000 mg/kg, YM exhibits strong tolerance (tolerance index greater than 90%), with normal growth and no signs of toxicity. However, SM showed a tolerance level of < 50% at high Pb concentrations, with significant heavy metal toxicity symptoms in the above-ground areas. These results provide important information for the remediation of Pb contaminated soils in mining areas.

Keywords: industrial hemp; lead (Pb); tolerance differences; lead accumulation

1. Introduction

In many terrestrial and aquatic ecosystems, the issue of heavy metal stress has become a major global concern. Due to the heavy metal accumulation, extensive industrialization in the recent past has been shown to have negative effects on agricultural productivity and soil quality [1]. Heavy metal, like lead (Pb), mercury (Hg), cadmium (Cd), arsenic (As), chromium (Cr), and nickel (Ni), have harmful effects on agricultural ecosystems. Due to their non-biodegradable features, they become a threat factor for human health and soil ecosystems [2,3]. Lead, as a heavy metal element typically toxic to plants, enters animal and plant systems through atmospheric deposition and soil absorption [3–6]. In plants, lead absorption disrupts critical physiological processes such as seed germination, root growth, and chlorophyll synthesis [7,8]. For instance, Pb was absorbed by the roots of hemp, and accumulated in the grains of the rice [9,10]. Pb lowers the photosynthetic rate of plants by damaging chloroplast ultrastructure, reducing chlorophyll production, blocking electron transport, and suppressing Calvin cycle enzyme activity [11]. Pb stress interferes with the plant's normal physiological processes. It suppresses enzyme activity, water balance, mineral nutrition, photosynthesis, and may lead to cell death [12]. However, the extent of lead accumulation, tolerance, and inhibition varies between plant species and even among varieties of the same species.

Cannabis sativa L. is an annual herbaceous plant of the *Cannabaceae* family, traditionally cultivated as an economic crop [13]. It is widely used in industrial, agricultural, and pharmaceutical sectors. Industrial hemp, a genetically improved variant with tetrahydrocannabinol (THC) levels below 0.3%, is distinguished by its short growth cycle, high biomass production, robust root system, adaptability, and ability to accumulate various heavy metals [9,14]. These characteristics not only confer high economic value but also highlight its significant potential for remediating heavy metal-contaminated soils [15].

The taproot of industrial hemp absorbs a large amount of non-nutrient elements from the soil during growth [16–18]. For example, studies by Golia et al. and Xu et al. demonstrated the strong Pb accumulation capacity of industrial hemp, making it uniquely advantageous for remediating heavy metal-contaminated soils [19,20]. Differences in Pb accumulation traits and tolerance mechanisms among various industrial hemp cultivars have not, however, been thoroughly studied.

Industrial hemp is widely recognized as an excellent potential crop for the remediation of heavy metal contaminated soils. However, previous studies have focused primarily on on-site remediation of field contamination. For example, Luyckx et al. (2022) replaced industrial hemp with bioenergy in lead-contaminated farmland in northern France by growing crops and spraying silicon on the pages to increase the heavy metal content of the crops and to improve and maintain the mechanical properties of the fibers [21]. Flajšman et al. (2023) investigated the uptake of heavy metals, such as Pb, by two industrial hemp varieties in areas with different levels of contamination, and found that the plant uptake of lead was highest in heavily contaminated soils [22]. The soils in these studies contained multiple heavy metals at the same time, and the natural environment was subject to a variety of uncontrollable factors that affected the experimental results. In view of this, this study focuses on two industrial hemp varieties with differing Pb tolerance levels. These two varieties originate from southern (YM) and northern (SM) part of China. This study examines the impact of Pb on the growth of these two species using pot experiments with artificially administered Pb to resemble natural environmental Pb stress. The study also looks at how different plant components absorb and accumulate Pb. Investigating the mechanisms of industrial hemp's tolerance to Pb stress and providing theoretical insights into its safe application in Pb-affected farmlands, soil restoration, and environmental enhancement are the objectives.

2. Materials and Methods

2.1. Experimental Materials

Two industrial hemp materials were used to study differences in Pb tolerance: Yunma 1 (YM), a Pb-tolerant, high-biomass, late-maturing variety bred by the Economic Crops Research Institute of the Yunnan Academy of Agricultural Sciences, and Shaanxi Industrial Hemp ym478 (SM), a Pb-sensitive, low-biomass, early-maturing variety. The test soil is red loam, derived from the cultivated layer (0–20 cm) at the experimental base of the Industrial Crops Research Institute of Yunnan Academy of Agricultural Sciences, Panlong

District, Kunming City, Yunnan Province, China. The physicochemical properties of soil are presented by Inductively Coupled Plasma Mass Spectrometry (ICAP RQ, Thermo Fisher Scientific, Waltham, MA, USA) (Table 1).

Table 1. Physical and chemical properties of the tested soil.

pH	Organic Matter (g/kg)	Total Nitrogen (N, %)	Total Phosphorus (P, %)	Total Potassium (K, %)	Hydrolyzable Nitrogen (N, mg/kg)	Available Phosphorus (P, mg/kg)	Available Potassium (K, mg/kg)	Total Lead (Pb, mg/kg)
6.69	97.5	0.392	0.213	0.176	263	27.6	642	65.03

2.2. Experimental Design

The pot experiment was carried out in the greenhouse of the Industrial Crops Research Institute of Yunnan Academy of Agricultural Sciences. Analytical grade Pb $(NO_3)_2$ was used as the treatment agent. Seven Pb^{2+} concentration levels were established, corresponding to soil Pb concentrations of 0 (65), 500 (647), 1000 (1468), 2000 (2497), 3000 (3275), 4000 (4861), and 5000 (5676) mg/kg. There were three "replicated" treatments. Each copy was a pot with a diameter of 50 cm, a height of 30 cm, and contained 21 kg of soil. After the addition of Pb, the soil was equilibrated for 15 days by 3 saturation cycles with distilled water and dry air. pot was given 2.35 g Potassium, 6.18 g Superphosphate, and 3.82 g Urea Sulfate as the base fertilizer. 20 seeds were sowed in each pot and 5 seedlings were kept after germination. After sowing, they were watered once a week, using 5 L of tap water per pot. Industrial hemp is grown at 18–25° with 13 h of sunlight in the day and 75% relative humidity. When the industrial hemp grows to the 3–4 pairs of true leaves stage, each pot is given 10 g of urea as a follow-up fertilizer. YM and SM plants reached seed maturity at 180 and 150 days, respectively. Plants were hand-harvested whole, dried naturally. Five plants per pot were sampled for traits, biomass, and metal content. All measurements were repeated three times. The roots of harvested plant samples were washed three times with tap water, then three to four times with distilled water, and finally vacuumed with absorbent paper. Each sample was re-watered to avoid cross-contamination of roots, stems, and leaves with soil.

2.3. Experimental Methods

2.3.1. Measurement of Agronomic Traits and Biomass of Two Industrial Hemp Varieties

A vernier caliper was used to measure stem diameter at harvest, and a measuring tape was used to measure plant height and root length. The biomass of the various plant parts was ascertained by measuring the dry weights of roots, stems, leaves, seeds, and fibers after the harvested plants were allowed to air dry naturally.

2.3.2. Determination of Pb Content in Samples

The dried plant samples were quenched at 105 °C for 0.5 h and dried in an oven at 70 °C to a constant weight, ground in an agate mortar with liquid nitrogen, and passed through a 0.42 mm sieve. They were weighed at approximately 0.2–0.5 g (accurate to 0.0001 g) with an electronic balance, then 5–10 mL of 68% nitric acid was added, they were covered and left for 1 h or overnight, according to the microwave digestion instrument (GB 5009.268-2016 National Standard for Food Safety Determination of Multi-Elements in Food Table B.1) operation. After cooling, ultrapure water of a volume up to 25 mL was used, then a blank test was performed.

ICP-MS (ICAP RQ, Thermo Fisher Scientific, Waltham, MA, USA) inductively coupled plasma mass spectrometry (GB-5009.268-2016 National Standard for Food Safety

Determination of Multiple Elements in Foods First Method) was used to determine the lead content.

2.3.3. Calculation of Accumulation and Tolerance Capacity in Industrial Hemp

The bioconcentration factor (BCF), which measures how well plants absorb heavy metals, was used to evaluate the plants' ability to accumulate heavy metals. It was found that the plant's capacity to absorb heavy metals was directly proportional to their bioconcentration factor (BCF) values [23,24]. The effectiveness of moving heavy metals from the roots to the above-ground organs is gauged by the translocation factor (TF), whose value shows how the metals are distributed and accumulate inside the plant.

The tolerance index (TI) is a key indicator of a plant's resistance to heavy metals, directly reflecting plant physiological adaptation to heavy metal stress. According to Lux et al., plants can be classified into three categories based on their TI values: highly tolerant (TI > 60%), moderately tolerant ($35\% \leq TI \leq 60\%$), and sensitive (TI < 35%) [25]. It is further noted that when plant biomass is reduced to 50% of the control group (i.e., TI = 50%), this can be considered the toxicity threshold for heavy metals, marking the critical point at which a plant transitions from tolerance to sensitivity [26]. Bioconcentration factor (BCF) = heavy metal content in a specific plant organ/heavy metal content in rhizosphere soil, translocation factor (TF) = heavy metal content in a specific plant organ/heavy metal content in the transport organ of the plant, and tolerance index (TI) = (Biomass of industrial hemp under Pb treatment/Biomass of control group industrial hemp) × 100%.

2.4. Data Statistical Analysis

SPSS 20.0 was used to analyze the experimental data, and GraphPad Prism 8 was used to visualize the results.

3. Results

3.1. The Effect of Pb Stress on Agronomic Traits of Industrial Hemp

The normal growth of plants can be indicated by their agronomic characteristics. In Pb-contaminated soils, the most directly impacted part is the root system, and root length, to some extent, reflects the development, growth, and metabolic status of the root system. After Pb is absorbed by roots, it will affect the plant's height and stem diameter; therefore, plant height, stem diameter, and root length are key indicators of the growth vigor of industrial hemp.

The effects of Pb stress on the growth of the two industrial hemp varieties are shown in Table 2. With an increase in Pb concentration, the plant height, stem diameter, and root length of both varieties initially increased and then decreased. Low concentrations of Pb exhibited a growth-promoting effect, whereas higher concentrations had an inhibitory effect on plant height and stem diameter. For YM, plants demonstrated strong growth within a Pb concentration range of 0–4000 mg/kg. However, when the concentration increased to 5000 mg/kg, plant growth was moderately inhibited. Despite this, the overall growth process continued to be normal. Compared to the control group, plant height decreased by roughly 15–17 cm, with notable changes seen as relative to other treatments. In addition, both stem diameter and root length decreased. At the seed maturity stage, no symptoms of heavy metal toxicity or plant death were observed, indicating that YM exhibited strong self-repair capacity. SM, the plant height, stem diameter, and root length followed a similar trend of initially increasing and then decreasing with increasing Pb concentration, peaking at a Pb concentration of 1000 mg/kg. When the Pb concentration reached 4000 mg/kg, these growth parameters were significantly inhibited, showing notable differences compared to the control group (CK). At a Pb dose of 5000 mg/kg, the height, stem diameter, and root length of SM decreased by 12.16%, 13.99%, and 13.01%, respectively. Compared with SM, YM has superior resistance to Pb stress.

Table 2. Agronomic traits of different hemp cultivars under Pb stress.

Varieties	Pb (mg/kg)	Plant Height (cm)	Stem Diameter (mm)	Root Length (cm)
YM	0	310.3 ± 2.1 b	10.2 ± 0.2 b	28.3 ± 0.5 a
	500	328.3 ± 5.9 a	12.5 ± 0.4 a	28.7 ± 0.6 a
	1000	308.1 ± 4.2 bc	10.0 ± 0.1 bcd	27.3 ± 0.3 b
	2000	303.2 ± 2.0 cd	10.3 ± 0.7 bc	26.6 ± 0.3 b
	3000	302.0 ± 3.0 d	9.7 ± 0.3 cd	25.8 ± 0.2 c
	4000	295.7 ± 4.0 e	9.5 ± 0.5 d	25.6 ± 0.5 c
	5000	293.0 ± 3.0 e	9.4 ± 0.5 d	22.9 ± 0.3 d
SM	0	148.7 ± 6.1 ab	6.8 ± 0.3 b	14.0 ± 0.5 ab
	500	149.7 ± 4.8 ab	7.0 ± 0.1 b	14.3 ± 0.5 ab
	1000	155.5 ± 5.3 a	7.6 ± 0.2 a	14.6 ± 0.4 a
	2000	147.1 ± 7.4 ab	6.7 ± 0.2 b	13.8 ± 0.4 b
	3000	145.9 ± 5.0 ab	6.8 ± 0.1 b	13.8 ± 0.3 b
	4000	135.5 ± 3.6 c	6.2 ± 0.2 c	12.6 ± 0.1 c
	5000	130.7 ± 4.4 c	5.8 ± 0.2 c	12.2 ± 0.5 c

Note: Data in the table are mean ± standard, different lowercase letters indicate significant differences between different concentrations ($p < 0.05$).

3.2. Effects of Different Pb Concentrations on the Biomass of Industrial Hemp Organs

The inhibitory effects of Pb stress on the biomass of different industrial hemp organs varied with Pb concentrations (Figure 1). At a low Pb concentration (500 mg/kg), the biomass of YM and SM organs was slightly higher than that of the control group, indicating that low Pb concentrations may have a stimulating impact on industrial hemp development of industrial hemp. With increase in Pb concentration, the biomass of industrial hemp organs demonstrated a decreasing tendency, which was significantly different from the control group.

For root biomass (Figure 1A) in the Pb concentration range of 500–1000 mg/kg, there was no significant difference between the two varieties and the control group. As the Pb content increased, the root biomass of both types gradually decreased. At Pb concentrations of 2000–4000 mg/kg for YM and 2000–3000 mg/kg for SM, significant differences ($p < 0.05$) were observed in root biomass compared to the control. However, the downward trend was not dramatic. However, under Pb stress at 5000 mg/kg for YM and 4000 mg/kg for SM, root growth was significantly inhibited. At these concentrations, root biomass decreased by 21.56% for YM and 17.47% for SM compared to control. The root biomass of YM was more than twice that of SM.

For stem and leaf biomass (Figure 1B,C), as the Pb concentration of the stems increased from 500 to 4000 mg/kg, YM biomass showed a statistically significant downward trend ($p < 0.05$), although the reduction was not significant. At a concentration of 5000 mg/kg, stem biomass was significantly impacted, showing a reduction of 16.45% compared to the control group. In SM, stem biomass showed no significant variation from control at Pb doses ranging from 500 to 2000 mg/kg. At concentrations above 3000 mg/kg, significant differences ($p < 0.05$) were observed compared to control. For leaves, low Pb concentrations (500–1000 mg/kg) did not significantly affect biomass compared to the control. At doses above 2000 mg/kg, Pb significantly impeded leaf growth, affecting biomass accumulation. At 5000 mg/kg, leaf biomass decreased by 44.67% relative to control, presumably due to Pb-induced cellular damage, which compromised photosynthetic efficiency. For fiber and seed biomass (Figure 1D,E), Pb stress had different effects on the accumulation of fiber and seed dry matter in YM and SM, especially at high Pb concentrations (5000 mg/kg). Compared with the control, the biomass of YM fibers and seeds decreased by 11.3% and 25.0%, respectively, but SM showed reductions of 31.1% and 31.7%, indicating a substantial inhibition of SM fibers and seeds. Compared to SM, YM showed greater resilience under elevated Pb stress. This was evident from its reduced biomass reduction across multiple organs, especially roots, fibers, and seeds.

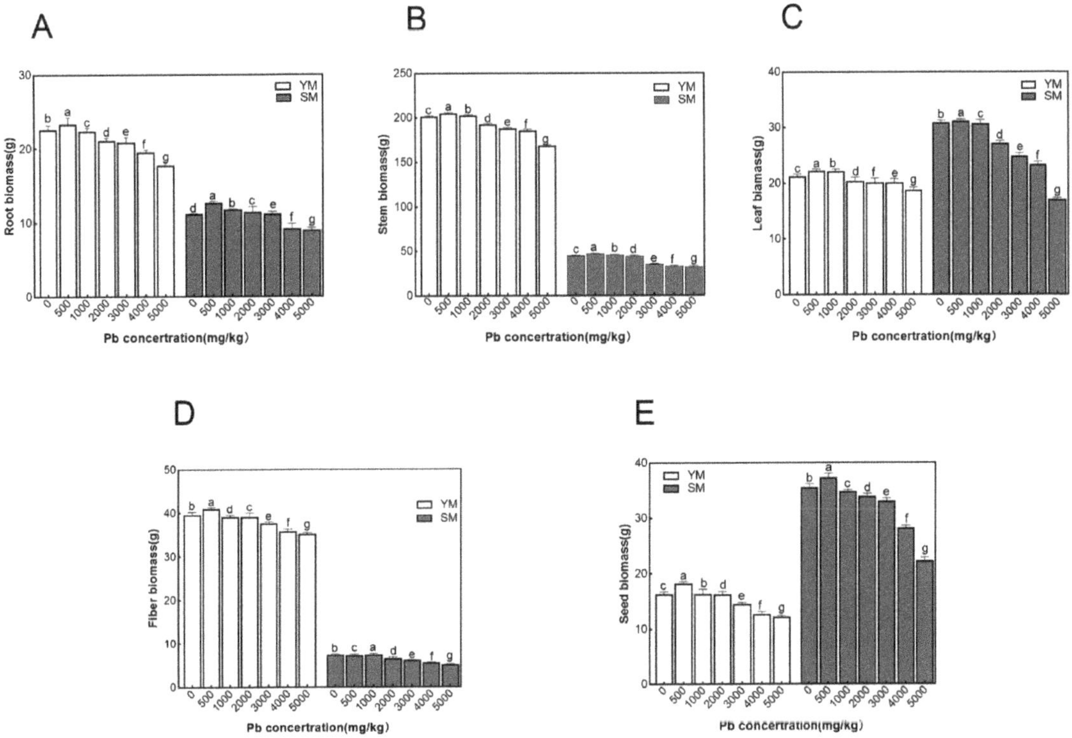

Figure 1. Biomass of different cultivars under Pb stress. (**A**): Comparison of root biomass; (**B**): Comparison of stem biomass; (**C**): Comparison of leaf biomass; (**D**): Comparison of fiber biomass; (**E**): Comparison of seed biomass. Data in the figure are mean ± standard, different lowercase letters indicate significant differences between different concentrations ($p < 0.05$).

3.3. Accumulation Characteristics of Pb in Two Industrial Hemp Varieties

As the Pb treatment concentration increased, the Pb content in all plant organs also increased, although the rate of increase varied significantly, with roots showing a much greater increase than other organs (Figure 2). Pb accumulation in different organs followed the pattern: roots > stems > leaves ≈ fibers > seeds. Lead was mainly contained in roots, followed by stems, leaves, and fibers, with the least amount found in seeds. However, the deposition of Pb varied significantly between the two materials. In YM, the concentration of Pb in roots was 1–20 times greater than in stems, leaves, and fibers, and 5–70 times higher than in seeds. In SM, when Pb concentration increased, Pb accumulation in roots progressively migrated to aerial organs, leading to elevated Pb levels in stems, leaves, and fibers. In SM, the Pb concentration in roots was only 0.17–0.43 times than that in stems, 1–10 times than that in leaves and fibers, and 8–170 times than that in seeds. In both materials, the lead level in the seeds was minimal; even at a lead concentration of 5000 mg/kg, the lead content in the seeds did not exceed 10 mg/kg. At soil Pb values below

2000 mg/kg, no significant variation in root Pb content was observed between YM and SM. However, when the concentration exceeded 3000 mg/kg, the Pb level in YM roots was significantly higher than in SM roots. At a dose of 5000 mg/kg, the lead content in YM roots was approximately twice that of SM roots, measuring 414.33 mg/kg and 242.70 mg/kg, respectively (Figure 2A). Figure 2B–D illustrate that at low Pb concentrations (0–500 mg/kg), Pb levels in stems, leaves, and fibers of both materials showed no significant differences. At values of 1000 mg/kg and above, the Pb content in YM was lower than that in SM. At 5000 mg/kg, Pb levels in the SM stems, leaves, and fibers were 1.34, 1.06, and 1.96 times higher than YM, respectively. In seeds (Figure 2E), the Pb concentration in both YM and SM was minimal (<10 mg/kg), although the Pb concentration in YM was significantly higher than in SM.

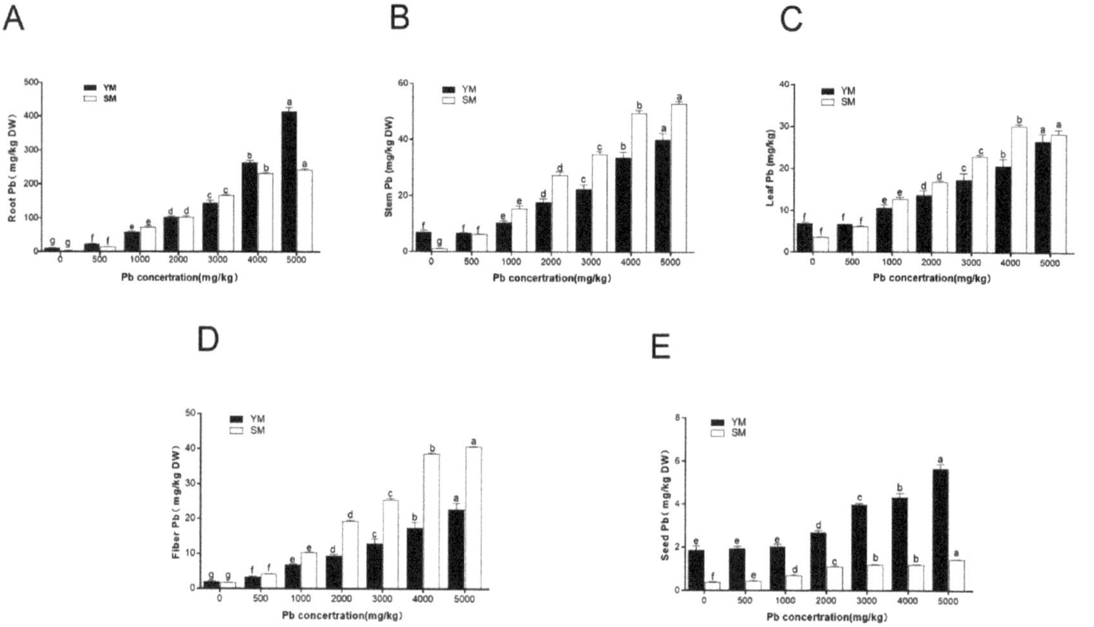

Figure 2. Pb content in different organs of two industrial hemp varieties under Pb stress (mg/kg). (**A**): Comparison of Pb content in roots; (**B**): Comparison of Pb content in stems; (**C**): Comparison of Pb content in leaves; (**D**): Comparison of Pb content in seeds; (**E**): Comparison of Pb content in fibers. Different lowercase letters represent significant differences in different concentrations ($p < 0.05$).

3.4. *Effects of Pb Stress on the Accumulation Capacity of Different Parts of the Two Industrial Hemp Varieties*

As shown in Figure 3, significant differences were observed in bioconcentration factors (BCFs) for roots, stems, leaves, fibers, and seeds of the two industrial hemp varieties. BCFs in roots were consistently higher than those in other organs. For both YM and SM, root BCFs decreased gradually as Pb concentration increased, with significant differences observed between treatments ($p < 0.05$). YM root BCF was higher than SM, especially at Pb concentrations of 0 mg/kg and 5000 mg/kg, where YM root BCF was 2.99 and 1.71 times that of SM, respectively.

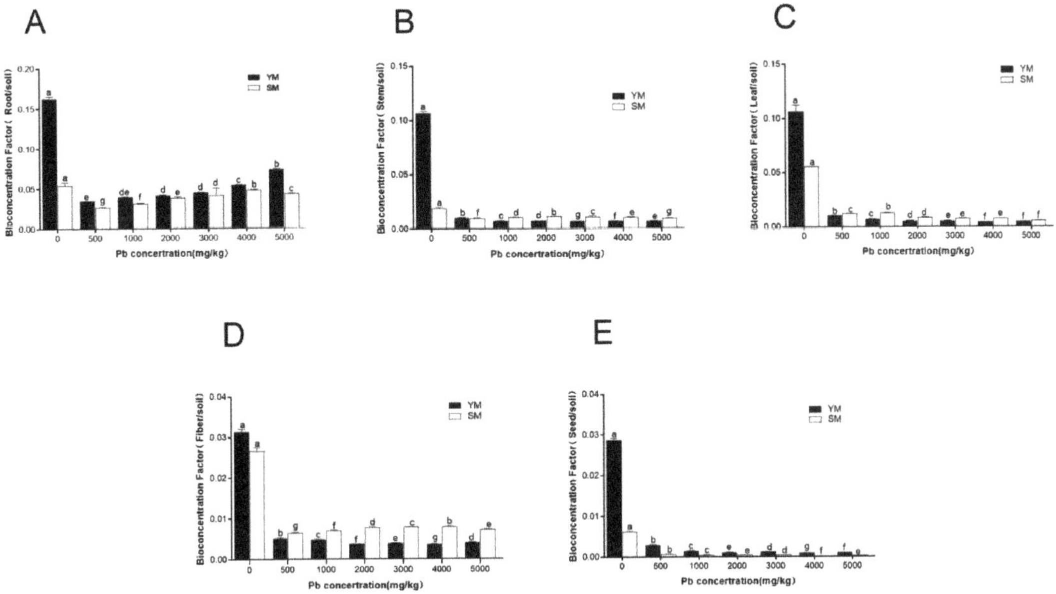

Figure 3. Bioconcentration factor of Pb from different organs in two industrial hemp varieties. (**A**): Root bioconcentration factor; (**B**): Stem bioconcentration factor; (**C**): Leaf bioconcentration factor; (**D**): Seed bioconcentration factor; (**E**): Fiber bioconcentration factor. Different lowercase letters represent significant differences in different concentrations ($p < 0.05$).

Figure 3B–D illustrate that BCFs in stems, leaves, and fibers for both YM and SM decreased to different extents as Pb concentration increased, with statistically significant differences between treatments ($p < 0.05$). In contrast to roots, YM BCFs in stems, leaves, and fibers were inferior to SM, except for the control group. Figure 3E shows that bioaccumulation factors (BCFs) for seeds in both YM and SM were low (BCF < 0.03). However, YM seed BCF was significantly higher than SM seed. While Pb mainly accumulated in the roots of both types, their bioconcentration capacities significantly. YM exhibited the ability to absorb significant amounts of Pb and withstand elevated Pb concentrations (5000 mg/kg). The bioconcentration capacity of SM decreased dramatically at a Pb concentration of 5000 mg/kg. YM showed significantly greater bioconcentration capacity in roots and seeds compared to SM, but SM exhibited superior BCFs in stems, leaves, and fibers compared to YM.

3.5. Effects of Pb Stress on the Translocation Ability of Different Parts of Two Industrial Hemp Varieties

Figure 4 demonstrates that Pb transfer coefficients (TFs) for stems, leaves, fibers, and seeds in both types of industrial hemp decreased progressively with elevated Pb concentrations, with significant variations seen between treatments ($p < 0.05$). Transcription factors for SM stems (excluding the control group) and leaves were elevated compared to YM (Figure 4A,B). This suggests that SM has a superior ability to translocate Pb from roots to aerial stems and leaves compared to YM. This may explain why SM leaves exhibited chlorosis and biomass loss at a Pb concentration of 4000 mg/kg, indicative of toxic symptoms. TFs for both YM and SM fibers decreased gradually as Pb concentration increased, with significant differences between treatments ($p < 0.05$). Although TFs for both varieties were generally low (TF < 0.6), TF for fibers were significantly lower than SM fibers

(Figure 4C). This suggests that a large amount of Pb transferred from the roots to stems in SM accumulated in bast fibers, causing severe damage to stem tissue.

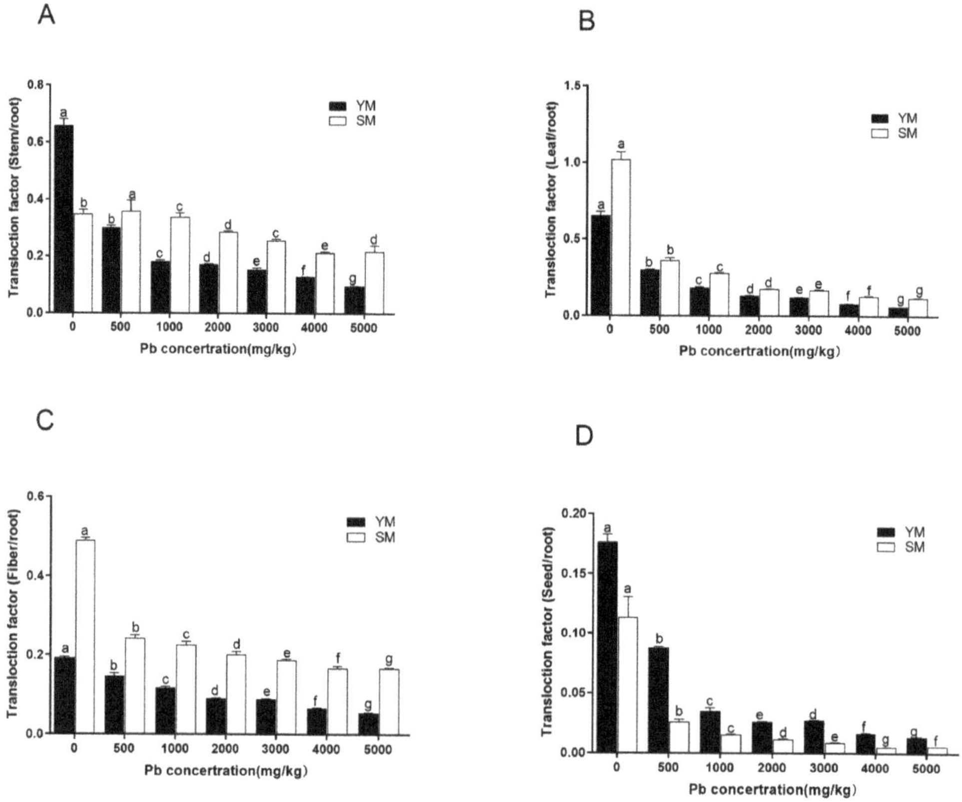

Figure 4. Translational factors of Pb form different organs in two industrial hemp varieties. (**A**): Stem translocation factor; (**B**): Leaf translocation factor; (**C**): Fiber translocation factor; (**D**): Seed translocation factor. Different lowercase letters represent significant differences in different concentrations ($p < 0.05$).

Correspondingly, seed transfer factors for both YM and SM decreased as Pb levels increased with significant differences between treatments ($p < 0.05$). Transcription factors (TFs) for seeds in both types were low (TF < 0.2), although YM exhibited significantly stronger TFs in seeds than SM (Figure 4D). In summary, SM exhibited a significantly superior ability to translocate Pb to stem, leaf, and fiber relative to YM, which likely resulted in increased damage observed in SM under elevated Pb stress. In contrast, YM had a superior ability to transfer Pb to seed relative to SM.

3.6. Comparison of Pb Tolerance Indices in Two Industrial Hemp Varieties

As the concentration of Pb increased, the root tolerance index (TI) initially increased and then decreased, with variations between the two varieties (Figure 5A). At a low Pb concentration (500 mg/kg), there was little difference in YM and SM root TIs, and both were higher than the control. Between 1000 and 3000 mg/kg, the TI for YM roots was lower than for SM. At concentrations of 4000–5000 mg/kg, both varieties exhibited root TIs above 60%, with YM showing slightly higher TIs than SM. This indicates that the roots of both varieties have a high tolerance to heavy metal Pb.

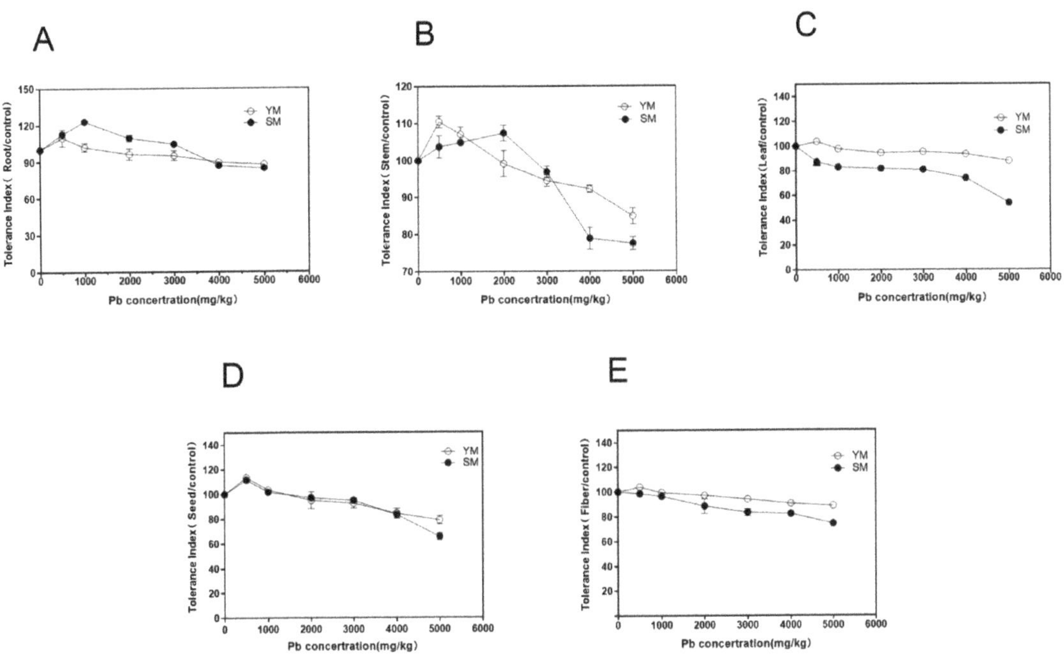

Figure 5. Tolerance index in different organs of two industrial hemp varieties under Pb stress. (**A**): Comparison of root tolerance index; (**B**): Comparison of stem tolerance index; (**C**): Comparison of leaf tolerance index; (**D**): Comparison of seed tolerance index; (**E**): Comparison of fiber tolerance index. The error line in the figure represents the mean ± standard.

For stems, the TI for YM was higher than the control at a low Pb concentration (500 mg/kg). As the concentration of Pb increased, TI gradually decreased, although the decrease was small, and TI remained above 60% across all concentrations (Figure 5B). In SM, stem TI increased as Pb concentration increased to 2000 mg/kg but began to decrease at concentrations above 2000 mg/kg. At 4000–5000 mg/kg, TI decreased to its lowest level and below YM, indicating that SM stems have a lower tolerance to Pb in high concentration environments compared to YM.

For leaves, YM showed a slow increase in TI as Pb concentration rose to 3000 mg/kg. At doses of 4000–5000 mg/kg, TI showed only a marginal decline, maintaining above 90% even at 5000 mg/kg, indicating significant Pb tolerance in YM leaves (Figure 5C). TI for SM leaves was below 50% at 5000 mg/kg. Wang et al. [26] indicates that a toxicity index of 50% serves as the threshold for heavy metal toxicity in plants. At a concentration of 5000 mg/kg, SM leaves showed a 50% decrease in biomass, indicating that the soil lead toxicity threshold for SM leaves is 5000 mg/kg. For seeds and fibers, YM showed higher TIs than control at a low Pb dose (500 mg/kg). As Pb concentration increased, TIs gradually decreased but remained above 60% at all concentrations (Figure 5D,E). However, the tolerance index for SM seeds at 5000 mg/kg was lower than that for YM seeds, suggesting that YM seeds exhibit more tolerance to Pb than SM seeds.

Overall, YM showed significant tolerance to Pb, with translocation indices for roots, stems, leaves, seeds, and fibers consistently exceeded 60% at all concentrations. In SM, TIs surpassed 60% at low Pb concentrations but decreased significantly in high Pb conditions. Notably, SM leaves showed a 50% TI at 5000 mg/kg, confirming the soil Pb toxicity threshold for SM leaves to be 5000 mg/kg. YM consistently shows superior tolerance to Pb in all plant organs relative to SM.

4. Discussion

4.1. Differences in Pb Tolerance Response Between Two Industrial Hemp Materials

Depending on the plant species and the specific organs involved, different plants have different capacities for absorbing and accumulating heavy metals [27–35]. For example, plants such as ramie, pepperwood, paper mulberry, and black nightshade mainly fix Pb in their roots, limiting their transport to parts above ground. This reduces damage to photosynthetic tissues and respiratory systems in above-ground organs, thereby enhancing Pb tolerance [36,37].

Pb is mostly accumulated in the roots of SM and YM cultivars in our investigation. Nevertheless, significant variations in Pb accumulation were noted between the two types. The Pb concentration in YM roots was twice that of SM roots under heavy Pb stress (>4000 mg/kg). YM showed that it could withstand elevated Pb concentrations and continue to grow normally under such circumstances. On the other hand, SM showed severe signs of heavy metal toxicity, including leaf chlorosis, poor seed development, hollow seeds, and premature aging, at comparable high Pb stress levels. The differences can be explained by the fact that YM retains a significant amount of lead (about 70%) in its roots, while only a small part (about 20%) is transferred to organs above ground. Pb-tolerant industrial hemp uses this retention mechanism as one of its methods to reduce heavy metal stress [38–40]. In contrast, SM caused high Pb concentrations in stems, leaves, and fibers by moving a large amount of Pb absorbed by the roots (>50%) to above-ground areas. SM may have a reduced tolerance to Pb due to the damage to terrestrial organs caused by this excessive accumulation, resulting in signs of Pb contamination. Although the bioconcentration factors (BCFs) and translocation factors (TFs) of both varieties did not meet the criteria for hyperaccumulator plants as proposed by Baker and Brooks [41] (BCF and TF > 1, Pb concentration in above-ground areas >1000 mg/kg, and no impact on normal plant growth under heavy metal stress), industrial hemp exhibited high biomass. In addition, YM showed no apparent signs of Pb toxicity under high Pb stress and showed higher tolerance and accumulation capacity than SM. Depending on local conditions, industrial hemp materials can be used selectively for the restoration of Pb contaminated soils in mining sites because of the different Pb accumulation and translocation capacities of both types.

4.2. Differences in Lead Tolerance Between Two Industrial Hemp Varieties

Plants' ability to acquire metals is only one aspect of their tolerance to heavy metals; another is their ability to continue growing normally in the face of heavy metal stress. As a key determinant of typical plant growth, biomass fully captures a plant's ability to store and use resources and energy [42,43]. According to Kandziora-Ciupa et al. [44], biomass allocation is a plant's method of balancing reproduction and survival while adjusting to its external environment. Increased biomass leads to a higher accumulation of heavy metals at constant concentrations. However, plant growth and development are impeded when heavy metals accumulate above a certain threshold. Arthur et al. [45] classified crops into high, medium, and low accumulators based on heavy metal accumulation levels. In this study, significant differences were observed between YM and SM in morphology, biomass, and tolerance to Pb stress.

1. Morphological differences: Plant height, stem diameter, and root length were somewhat suppressed in YM at Pb concentrations of 4000–5000 mg/kg, but no contamination symptoms were observed. In contrast, SM exhibited severe toxicity symptoms, including leaf chlorosis, poor seed development, hollow seeds, and premature aging. This phenomenon of "low promotion, high inhibition" under Pb stress is consistent

with findings in other plants, such as Sedum alfredii, where low Pb concentrations promote growth while high concentrations inhibit growth [46,47].

2. Biomass differences: YM biomass was more than 2.5 times larger than SM biomass. When soil Pb concentration was less than 1000 mg/kg, industrial hemp growth was promoted, with higher root and leaf biomass compared to the control group. However, biomass accumulation and plant development were inhibited at elevated Pb concentrations. Both YM and SM showed a considerable inhibition of root growth at Pb values of 5000 mg/kg and 4000 mg/kg. At doses as low as 3000 mg/kg, the biomass of SM differed significantly from control ($p < 0.05$), and at 5000 mg/kg, it decreased by more than 50%, suggesting a severe suppression of plant growth.
3. Tolerance differences: The tolerance indices (TIs) for roots and stems in both YM and SM exceeded 60%. However, at high Pb concentrations (>4000 mg/kg), TI for YM was significantly higher than for SM. In leaf tissue, YM exhibited higher TIs than SM, with TI exceeding 90% even at 5000 mg/kg. However, at Pb concentrations above 5000 mg/kg, SM showed a TI < 50%, suggesting that SM leaves have a Pb toxicity threshold of 5000 mg/kg. Industrial hemp is classified as a medium accumulator with high tolerance based on the accumulation characteristics seen in YM and SM in this study as well as the criteria established by Arthur [45] for dividing crops into high, medium, and low accumulators of heavy metals. In addition, YM showed greater Pb tolerance than SM.

5. Conclusions

Industrial hemp, as a fiber crop not used for consumption, ensures that heavy metals do not enter the food chain, making it an excellent candidate for remediating heavy metal contaminated soils. Regarding Pb accumulation in tissues, Pb mainly accumulates in roots of both YM and SM varieties, with YM showing a higher accumulation capacity than SM. Only 30% of the Pb absorbed by YM was transferred to the above-ground portions; the remaining 70% was sequestered in its roots. On the other hand, under high Pb conditions, SM caused toxicity symptoms by translocating 50% of the Pb that had accumulated in the roots to its stems and leaves. Industrial hemp is classified as a medium accumulator with exceptional Pb tolerance by combining phenotypic, biomass, and Pb accumulation and translocation assessments. When it comes to Pb tolerance, YM is more resilient than SM.

Author Contributions: Conceptualization, Y.X.; methodology, Y.X.; investigation, Q.Z. and R.G.; data curation, A.K., Y.Z. and P.L.; writing—original draft preparation, Y.X. and A.K.; writing—review and editing, Y.X., X.A., X.C., Q.J. and M.Y.; visualization, P.L.; funding acquisition, Y.X. All authors have read and agreed to the published version of the manuscript.

Funding: This study was supported by the National Natural Science Foundation of China (32460460), Applied Basic Research Key Project of Yunnan (202401BD070001-015), and Yunnan Intelligence Union Program (YIUP) (202303AM140009).

Institutional Review Board Statement: Not applicable.

Informed Consent Statement: Not applicable.

Data Availability Statement: The data that support the findings of this study are available from the corresponding author upon reasonable request.

Conflicts of Interest: The authors declare no conflicts of interest.

References

1. Teiba, I.I.; El-Bilawy, E.H.; Abouelsaad, I.A.; Shehata, M.; Alhoshy, Y.J.; Habib, N.M.; Abu-Elala, N.; Belal, E.B.; El-Khateeb, W.A.M. The role of marine bacteria in modulating the environmental impact of heavy metals, microplastics, and pesticides: A comprehensive review. *Environ. Sci. Pollut. Res.* **2024**, *31*, 64419–64452. [CrossRef] [PubMed]
2. Ran, C.; Liu, C.; Peng, C.; Li, X.; Liu, Y.; Li, Y.; Zhang, W.; Cai, H.; Wang, L. Oxidative potential of heavy-metal contaminated soil reflects its ecological risk on earthworm. *Environ. Pollut.* **2023**, *323*, 121275. [CrossRef] [PubMed]
3. Vareda, J.P.; Valente, A.J.; Durães, L. Assessment of heavy metal pollution from anthropogenic activities and remediation strategies: A review. *J. Environ. Manag.* **2019**, *246*, 101–118. [CrossRef] [PubMed]
4. Liu, Z.; Zhang, Q.; Han, T.; Ding, Y.; Sun, J.; Wang, F.; Zhu, C. Heavy Metal Pollution in a Soil-Rice System in the Yangtze River Region of China. *Int. J. Environ. Res. Public Health* **2016**, *13*, 63. [CrossRef]
5. Ali, H.; Khan, E.; Ilahi, I. Environmental chemistry and ecotoxicology of hazardous heavy metals: Environmental persistence, toxicity, and bioaccumulation. *J. Chem.* **2019**, *2019*, 6730305. [CrossRef]
6. Ekmekyapar, F.; Sabudak, T.; Seren, G. Assessment of heavy metal contamination in soil and wheat (*Triticum aestivum* L.) plant around the Çorlu–Çerkezkoy highway in Thrace region. *Glob. NEST J.* **2012**, *14*, 496–504.
7. Ni'am, M.I.; Yuniati, R. Effect of lead (Pb) on seed germination of water spinach (*Ipomoea aquatica* Forsk). *J. Phys. Conf. Ser.* **2021**, *1725*, 012041. [CrossRef]
8. Marek, B. Influence of lead on the chlorophyll content and on initial steps of its synthesis in greening cucumber seedlings. *Acta Soc. Bot. Pol.* **1985**, *54*, 95–105.
9. Ahmad, R.; Tehsin, Z.; Malik, S.T.; Asad, S.A.; Shahzad, M.; Bilal, M.; Shah, M.M.; Khan, S.A. Phytoremediation potential of hemp (*Cannabis sativa* L.): Identification and characterization of heavy metals responsive genes. *CLEAN-Soil Air Water* **2016**, *44*, 195–201. [CrossRef]
10. Chang, J.D.; Gao, W.; Wang, P.; Zhao, F.J. Osnramp5 is a major transporter for lead uptake in rice. *Environ. Sci. Technol.* **2022**, *56*, 17481–17490. [CrossRef]
11. Zafar, S.; Khan, M.K.; Aslam, N.; Hasnain, Z. Impact of Different Stresses on Morphology, Physiology, and Biochemistry of Plants. In *Molecular Dynamics of Plant Stress and Its Management*; Shahid, M., Gaur, R., Eds.; Springer: Berlin/Heidelberg, Germany, 2024; pp. 67–91.
12. Shukla, S.; Das, S.; Phutela, S.; Triathi, A.; Kumari, C.; Shukla, S.K. Heavy metal stress in plants: Causes, impact and effective management. In *Heavy Metal Toxicity: Human Health Impact and Mitigation Strategies*; Kumar, N., Ed.; Environmental Science and Engineering; Springer Nature: Cham, Switzerland, 2024; pp. 187–215.
13. Krystyna, Z.G.; Janusz, G. *Cannabis sativa* L.–cultivation and quality of raw material. *J. Elementol.* **2018**, *23*, 971–984.
14. Rheay, H.T.; Omondi, E.C.; Brewer, C.E. Potential of hemp (*Cannabis sativa* L.) for paired phytoremediation and bioenergy production. *GCB Bioenergy* **2020**, *13*, 525–536. [CrossRef]
15. Linger, P.; Müssig, J.; Fischer, H.; Kobert, J. Industrial hemp (*Cannabis sativa* L.) growing on heavy metal contaminated soil: Fibre quality and phytoremediation potential. *Ind. Crops Prod.* **2002**, *16*, 33–42. [CrossRef]
16. Testa, G.; Corinzia, S.A.; Cosentino, S.L.; Ciaramella, B.R. Phytoremediation of Cadmium-, Lead-, and Nickel-Polluted Soils by Industrial Hemp. *Agronomy* **2023**, *13*, 995. [CrossRef]
17. Ferrarini, A.; Fracasso, A.; Spini, G.; Fornasier, F.; Taskin, E.; Fontanella, M.C.; Beone, G.M.; Amaducci, S.; Puglisi, E. Bioaugmented phytoremediation of metal-contaminated soils and sediments by hemp and giant reed. *Front. Microbiol.* **2021**, *12*, 645893. [CrossRef]
18. Todde, G.; Marras, S.; Caria, M.; Sirca, C.; Carboni, G. Industrial hemp (*Cannabis sativa* L.) for phytoremediation: Energy and environmental life cycle assessment of using contaminated biomass as an energy resource. *Sustain. Energy Technol. Assess* **2022**, *52*, 102081. [CrossRef]
19. Golia, E.E.; Bethanis, J.; Ntinopoulos, N.; Kaffe, G.G.; Komnou, A.A.; Vasilou, C. Investigating the potential of heavy metal accumulation from hemp. The use of industrial hemp (*Cannabis Sativa* L.) for phytoremediation of heavily and moderated polluted soils. *Sustain. Chem. Pharm.* **2022**, *31*, 100961. [CrossRef]
20. Xu, Y.P.; Deng, G.; Guo, H.Y.; Yang, M.; Yang, Q.H. Accumulation and sub cellular distribution of lead (Pb) in industrial hemp grown in Pb contaminated soil. *Ind. Crops Prod.* **2021**, *161*, 113220. [CrossRef]
21. Luyckx, M.; Blanquet, M.; Isenborghs, A.; Guerriero, G.; Bidar, G.; Waterlot, C.; Douay, F.; Lutts, S. Impact of silicon and heavy metals on hemp (*Cannabis sativa* L.) bast fibres properties: An industrial and agricultural perspective. *Int. J. Environ. Res.* **2022**, *16*, 82. [CrossRef]
22. Tlustoš, P.; Száková, J.; Hrubý, J.; Hartman, I.; Najmanová, J.; Nedělník, J.; Pavlíková, D.; Batysta, M. Removal of As, Cd, Pb, and Zn from contaminated soil by high biomass producing plants. *Plant Soil Environ.* **2006**, *52*, 413–423. [CrossRef]

23. Arnot, J.A.; Gobas, F.A.P.C. A review of bioconcentration factor (BCF) and bioaccumulation factor (BAF) assessments for organic chemicals in aquatic organisms. *Environ. Rev.* **2006**, *14*, 257–297. [CrossRef]
24. Ding, C.F.; Ma, Y.B.; Li, X.G.; Zhang, T.L.; Wang, X.X. Derivation of soil thresholds for lead applying species sensitivity distribution: A case study for root vegetables. *J. Hazard. Mater.* **2016**, *303*, 21–27. [CrossRef] [PubMed]
25. Lux, A.; Šottníková, A.; Opatrná, J.; Greger, M. Differences in structure of adventitious roots in Salix clones with contrasting characteristics of cadmium accumulation and sensitivity. *Physiol. Plant.* **2004**, *120*, 537–545. [CrossRef]
26. Wang, S.; Shi, X.; Sun, H.; Chen, Y.; Pan, H.; Yang, X.; Rafiq, T. Variations in metal tolerance and accumulation in three hydroponically cultivated varieties of *Salix integra* treated with lead. *PLoS ONE* **2014**, *9*, e108568. [CrossRef]
27. Karavin, N.; Batın, I. Resorption, mobilization and accumulation of metals in different parts of *Vitis vinifera* L. *Vitis* **2020**, *59*, 105–109.
28. Yang, K. Absorption and accumulation of heavy metals of Pb, Cu by different waterlilies. *Bot. Res.* **2019**, *8*, 366–369.
29. Mukherjee, S.; Singh, S.K.; Jatav, S.S.; Patra, A.; Reddy, G.P. Effect of biochar application on heavy metal accumulation in different parts of paddy plant. *Int. J. Environ. Clim. Change* **2023**, *13*, 4491–4500. [CrossRef]
30. Koubová, K.; Tlustoš, P.; Břendová, K.; Száková, J.; Najmanová, J. Lead accumulation ability of selected plants of *Noccaea* spp. *Soil Sediment Contam.* **2016**, *25*, 882–890. [CrossRef]
31. Lin, C.; Wang, Y.; Hu, G.; Yu, R.; Huang, H. Source apportionment and transfer characteristics of Pb in a soil-rice-human system, Jiulong River Basin, southeast China. *Environ. Pollut.* **2023**, *326*, 121489. [CrossRef]
32. Bech, J.; Duran, P.; Roca, N.; Poma, W.; Sánchez, I.; Roca-Pérez, L.; Boluda, R.; Barceló, J.; Poschenrieder, C. Accumulation of Pb and Zn in Bidens triplinervia and Senecio sp. spontaneous species from mine spoils in Peru and their potential use in phytoremediation. *J. Geochem. Explor.* **2012**, *123*, 109–113. [CrossRef]
33. Blanco, A.; Salazar, M.J.; Vergara Cid, C.; Pereyra, C.; Cavaglieri, L.R.; Becerra, A.G.; Pignata, M.L.; Rodriguez, J.H. Multidisciplinary study of chemical and biological factors related to Pb accumulation in sorghum crops grown in contaminated soils and their toxicological implications. *J. Geochem. Explor.* **2016**, *166*, 18–26. [CrossRef]
34. Salazar, M.J.; Rodriguez, J.H.; Cid, C.V.; Pignata, M.L. Auxin effects on Pb phytoextraction from polluted soils by Tegetes minuta L. and Bidens pilosa L.: Extractive power of their root exudates. *J. Hazard. Mater.* **2016**, *311*, 63–69. [CrossRef] [PubMed]
35. Htwe, W.M.; Kyawt, Y.Y.; Thaikua, S.; Imai, Y.; Mizumachi, S.; Kawamoto, Y. Effects of liming on dry biomass, lead concentration and accumulated amounts in roots and shoots of three tropical pasture grasses from lead contaminated acidic soils. *Grassl. Sci.* **2016**, *62*, 257–261. [CrossRef]
36. Li, S.M.; Wang, Q.Y.; Li, W.J.; Yang, Y.; Jiang, L.J. Litsea males are better adapted to Pb stress than females by modulating photosynthesis and Pb subcellular distribution. *Forests* **2023**, *14*, 724. [CrossRef]
37. Chen, J.; Han, Q.Q.; Duan, B.L.; Korpelainen, H.; Li, C.Y. Sex-specific competition differently regulates ecophysiological responses and phytoremediation of *Populus cathayana* under Pb stress. *Plant Soil* **2017**, *421*, 203–218. [CrossRef]
38. Hassan, S.H.; Chafik, Y.; Lebrun, M.; Sferra, G.; Bourgerie, S.; Scippa, G.S.; Morabito, D.; Trupiano, D. Lead toxicity and tolerance in plants. In *Heavy Metal Toxicity and Tolerance in Plants: A Biological, Omics, and Genetic Engineering Approach*; Hossain, M.A., Zakir Hossain, A.K.M., Bourgerie, S., Fujita, M., Dhankher, O.P., Haris, P., Eds.; Wiley: New York, NY, USA, 2023; pp. 373–405.
39. Amin, H.; Arain, B.A.; Jahangir, T.M.; Abbasi, A.R.; Abbasi, M.S.; Amin, F. Phytostabilzation as a sustainable phytoremediation strategy for lead contaminated soil–Screening of biofuel plants for lead tolerance and accumulation. *J. Plant Resour. Environ.* **2020**, *4*, 18.
40. Khaliq, S.; Iqbal, M.; Yaseen, W.; Rasheed, R. The exogenous menadiol diacetate enhances growth and yield by reducing Pb uptake, translocation and its toxicity through tissue nutrients acquisition in cucumber (*Cucumis sativus* L.). *Environ. Technol. Innov.* **2021**, *23*, 101666. [CrossRef]
41. Baker, A.J.M.; Brooks, R.R. Terrestrial higher plants which hyperaccumulate metallic elements. A review of their distribution, ecology and phytochemistry. *J. Ethnopharmacol.* **1989**, *1*, 81–126.
42. Chao, S.; Du, W.; Lu, T.; Yang, Y.; Wang, K.; Du, H.; Cheng, H.; Yu, D. Genome-wide association study of soybean (Glycine Max) phosphorus deficiency tolerance during the seedling stage. *Plant Breed.* **2021**, *140*, 267–284. [CrossRef]
43. Cai, X.Y.; Jiang, M.Y.; Liao, J.R.; Yang, Y.X.; Li, N.F.; Cheng, Q.B.; Li, X.; Song, H.X.; Luo, Z.H.; Liu, S.L. Biomass allocation strategies and Pb-enrichment characteristics of six dwarf bamboos under soil Pb stress. *Ecotoxicol. Environ. Saf.* **2020**, *207*, 111500. [CrossRef]
44. Kandziora-Ciupa, M.; Nadgórska-Socha, A.; Barczyk, G.; Ciepał, R. Bioaccumulation of heavy metals and ecophysiological responses to heavy metal stress in selected populations of *Vaccinium myrtillus* L. and *Vaccinium vitis-idaea* L. *Ecotoxicology* **2017**, *26*, 966–980. [CrossRef] [PubMed]
45. Arthur, E.; Crews, H.; Morgan, C. Optimizing plant genetic strategies for minimizing environmental contamination in the food chain. *Int. J. Phytoremediat.* **2000**, *2*, 1–21. [CrossRef]

46. Xiong, Y.H.; Yang, X.E.; Ye, Z.Q.; He, B. Comparing the characteristics of growth response and accumulation of cadmium and lead by *Sedum alfredii* Hance. *J. Northwest A F Univ. (Nat. Sci. Ed.)* **2004**, *32*, 101–106.
47. Ghavri, S.V.; Singh, R.P. Growth, biomass production and remediation of copper contamination by Jatropha curcas plant in industrial wasteland soil. *J. Environ. Biol.* **2012**, *33*, 207.

toxics MDPI

Article

Source Apportionment of Potentially Toxic Elements in Agricultural Soils of Yingtan City, Jiangxi Province, China: A Principal Component Analysis–Positive Matrix Factorization Method

Shaoting Chen [1,2], Hongmei Wang [1,*] and Ruiming Han [2]

1 Institute of Water Ecology and Environment Research, Chinese Research Academy of Environmental Sciences, Beijing 100012, China; 222502020@njnu.edu.cn
2 School of Environment, Nanjing Normal University, Nanjing 210023, China; 09386@njnu.edu.cn
* Correspondence: wanghmxj@163.com

Abstract: The increase in the concentration of potentially toxic elements in farmland soil attracts more and more attention. To identify the sources of potentially toxic elements in agricultural soils, 148 soil samples in Yingtan were selected as a case study, potentially toxic elements levels were analyzed, and principal component analysis (PCA) and positive matrix factorization (PMF) were employed. The results indicate that the average of Zn (89.62 mg· kg^{-1} d.w.), Cu (76.30 mg· kg^{-1} d.w.), Pb (35.56 mg· kg^{-1} d.w.), Mo (0.66 mg· kg^{-1} d.w.), and Cd (0.59 mg· kg^{-1} d.w.) exceed the corresponding soil background values of Jiangxi Province. Moreover, the high spatial coefficient of variation (above 1.00) for these elements suggests a significant influence from long-term external inputs. Among all of the elements above, the soil Cu and Cd concentrations indicate a relatively high pollution ranked by *Igeo*. Further analysis of sources apportioned by PCA and PMF implies that the potentially toxic elements input into agricultural soil may be attributed to mining activity, natural sources, smelting, and agricultural activity. This study implies that PCA-PMF combined with the field survey may be helpful tools for discerning the pollutants′ sources, and it addresses a view that the increasing Cu and Cd levels in farmland is concerning, as it is associated with the historical use of mixed fertilizers and a lack of supervision.

Keywords: agricultural soil; potentially toxic elements (PTEs); principal component analysis (PCA); positive matrix factorization (PMF)

1. Introduction

Over the past four decades, highly intensive agricultural production alongside rapid industrialization and urbanization has led to increasingly severe farmland soil pollution and quality decline in China, posing a grave threat to agrarian product quality and safety [1]. According to the National Soil Pollution Status Survey Bulletin [2], 16.1% of the soil samples and 19.4% of the agricultural soils are contaminated based on China's soil environmental quality limits, mainly by potentially toxic elements (PTEs), with cadmium pollution exceeding the standard the most severely (7.0%).

Currently, PTE pollution in China's farmland is characterized by an expanding area and increasingly diverse sources [3]. Agricultural soil pollution distribution is spatially uneven, with contamination being more severe in the south and east compared to the north and west. Regions such as the Yangtze River Delta, Pearl River Delta, and the Northeast old industrial bases, which are major grain-producing areas with intensive industrial activities and high agricultural intensification, face prominent soil issues [4]. The sources of PTEs are also relatively complex, with parent materials and human activities being the main factors affecting their sources. Human activities mainly include industrial production emissions, sewage irrigation, mineral resource development, pesticide and fertilizer use, etc. [5,6]. However, with the modernization of agriculture and the development of the rural economy, the scope and severity of PTE pollution are expanding and worsening [7]. The safe utilization of agricultural soils in China is gathering increasing concern [8,9].

Exceeding the contaminant limits in food crops is widespread in some areas, especially Cd in rice in the southern region, which is often related to contaminated agricultural soils [10,11]. There are various inputs of contaminants, such as anthropogenic activities involving irrigation water, atmospheric deposition, utilizing fertilizers containing potentially toxic elements, accumulating hazardous waste containing PTEs [12,13], and even leachate from hazardous waste disposal [14–16]. Except for PTEs due to human activities, the background PTEs in soil or its acidic nature are thought to promote crop species or cultivars prone to PTE accumulation. An accurate identification of contaminant sources in agricultural systems is crucial for risk control strategies.

Source apportionment techniques can be categorized into qualitative source identification and quantitative source contribution. Qualitative methods primarily use principal component analysis (PCA) and isotope fingerprint analysis to identify major sources, while quantitative methods employ receptor models such as positive matrix factorization (PMF), geostatistical analysis, and chemical mass balance (CMB) [17–20]. PCA and PMF are widely used for PTE source apportionment in soils [21–23]. PCA simplifies data and reveals intrinsic structures by reducing dimensionality, which enhances source identification efficiency while preserving original data information. PMF allocates sources without prior knowledge of source profiles, which is advantageous when profiles are unknown or difficult to obtain [24]. However, PCA is qualitative, while PMF is quantitative, lacking an intuitive display of factor numbers and requiring multiple calculations [25]. Combining PCA and PMF compensates for their shortcomings, providing precise estimates of each element's source contributions and minimizing biased judgments in source apportionment. For example, a study using this combined approach found that coal combustion (34.15%), livestock farming (17.44%), traffic emissions (12.42%), and natural factors (35.99%) were the main sources of pollution [26].

However, the complexity of the environment inevitably affects the accuracy of source resolution. If the sample number is limited, more variations may occur. Therefore, the combined auxiliary information should be obtained and integrated into PCA-PMF to improve the prediction's accuracy.

The primary aim of this study is to develop PCA-PMF techniques by combined land-use type, which will be used for the source apportionment of PTEs in agricultural soils. This study emphasizes the advantage of the improved PCA-PMF and sheds light on the source of the PTEs of farmland, which is expected to help with risk reduction strategies on agricultural soil.

2. Materials and Method

2.1. Study Area and Sampling

The study area is located in Yingtan City, Jiangxi Province, China. It is characterized by a subtropical, humid monsoon climate; the region experiences high temperatures, ample sunlight, and abundant rainfall. The predominant soil type is Ferralsols (based on the FAO Soil Classification System), and the primary crop cultivated is rice. Yingtan City is endowed with abundant mineral resources, with reserves of Pb, Zn, Ag, and Mn ranking among the top in Jiangxi Province. The Guixi Smelter, the largest modern copper smelter in China, is located in the research area. Its main products include copper, sulfuric acid, electrolytic copper, electrolytic silver, etc. In 2022, a total of 148 agricultural soil samples (depth of 0–20 cm) were collected using clean stainless-steel shovels (Delixi Electric Co., Ltd., Leqing, China), each comprising a mixture of five closely located subsamples. After applying the quartering method, approximately 1 kg of the mixed soil sample was retained and stored in a polyethylene bag away from light, and the sampling information was recorded. The sampling sites are shown in Figure 1.

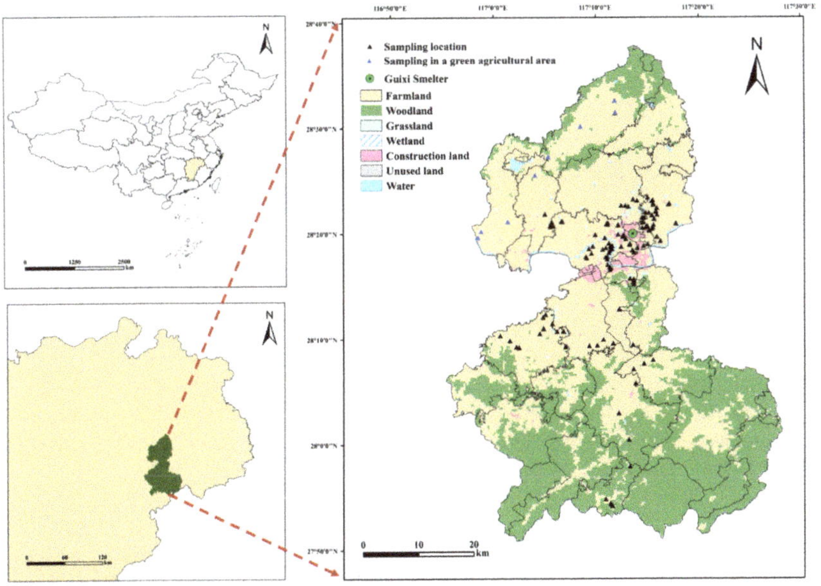

Figure 1. Sample sites.

2.2. Field Investigation

Zhoufang Town (ZF Town) is situated at the northern end of Guixi City, Yingtan City, approximately 30 km from the Guixi Smelter. It is a typical green agricultural area with no industrial activities, sharing a similar climate, geological background, and agricultural cultivation situation with the aforementioned research area. According to the local data regarding the safe use and stringent control of contaminated agricultural land in 2021, certain rice samples have been identified as having PTE levels that exceed the limits set by the National Food Safety Standard (GB 2762-2012) [27].

Field investigations have been carried out along with face-to-face interviews to obtain more historical information on the farmland. The data reveal that the main irrigation water is taken from the local reservoir storage or a small river nearby. According to the villagers' statements, it was confirmed that for some land, Cd rice exceeded the standard and had a history of mixed fertilizers being applied. The compound fertilizers had a lack of labels relating to PTE testing and were purchased from various sellers and producer sites, including Jiangxi, Hunan, Hubei, and Zhejiang. The usage of compound fertilizers is estimated to reach approximately 50~80% of the total fertilizer, but no detailed purchase and usage data have been recorded. Some sites are reported as old farmhouses or woodland; thus, they are assigned as soil without any fertilizer added. In some areas, slag piles in the surface soil have been observed along rural roads with low traffic volume, far away from rush-hour highways. The sample sites were classified based on the field investigation.

2.3. Analysis of PTEs

The soil samples were first sieved to remove debris such as roots, leaves, and stones. They were then dried at 60 °C to a constant weight for 48 h to ensure the removal of all free water. After drying, the samples were ground and passed through a 100-mesh sieve to obtain a homogeneous powder. All subsequent analyses were conducted on these prepared samples, and the results are reported on a dry weight basis (mg· kg^{-1} d.w.). The samples were comprehensively analyzed at the China Customs Science and Technology Research Center (STRC), which holds the necessary certification qualifications from China Metrology Accreditation (CMA) and is recognized as a proficiency testing provider by the China National Accreditation Service for Conformity Assessment (CNAS PT0012). The sample preparation and analysis process strictly adhered to the soil and sediment determination of aqua regia extracts of 12 metal elements by inductively coupled plasma mass spectrometry (HJ 803-2016) [28].

The determination of PTEs was carried out on an ICP-MS (PerkinElmer NexION 350X, Waltham, MA, USA) with a carrier gas flow rate of 1.10 L/min and a sampling depth of 6.9 mm. The instrument automatically measured three times. Two laboratory blank samples per batch were analyzed, with results below the detection limit. A zero-point analysis of the standard curve was conducted after each batch analysis. Furthermore, parallel sample analyses and spiked recovery tests were conducted on 10% of the samples per batch. The detection results met the standard requirements (refer to HJ 803-2016), with a relative standard deviation of less than 5% and a recovery rate ranging from 95.6% to 104.0%. A total of twelve PTEs were detected, including vanadium (V), chromium (Cr), manganese (Mn), cobalt (Co), nickel (Ni), copper (Cu), zinc (Zn), arsenic (As), molybdenum (Mo), cadmium (Cd), antimony (Sb), and lead (Pb).

2.4. Index of Geo-Accumulation

The Index of Geo-accumulation (*Igeo*), as introduced by Müller, is a widely recognized tool for assessing the degree of potentially toxic element contamination. It compares the element concentrations in the sampled soils to those in the background environment [29,30]. It is calculated using Equation (1).

$$I_{geo} = \log_2[C_n/\mathrm{k} \times B_n] \tag{1}$$

Here, C_n (mg· kg^{-1} d.w.) represents the measured PTE content in sampled soil, while B_n denotes the background value of potentially toxic elements in soil (mg· kg^{-1} d.w.) based on the soil background values in Jiangxi Province. k, usually denoted as 1.5, is the correction factor that indicates the possible variation of the background value of potentially toxic elements in the soil rock layer [30–33]. The *Igeo* values are categorized into seven classes, ranging from 0 to 6 (Table 1).

Table 1. I_{geo} of contamination degrees ranking.

Class	I_{geo} Value	Soil Quality
0	$I_{geo} \leq 0$	Practically uncontaminated
1	$0 < I_{geo} \leq 1$	Uncontaminated to moderately contaminated
2	$1 < I_{geo} \leq 2$	Moderately contaminated
3	$2 < I_{geo} \leq 3$	Moderately to heavily contaminated
4	$3 < I_{geo} \leq 4$	Heavily contaminated
5	$4 < I_{geo} \leq 5$	Heavily to extremely contaminated
6	$I_{geo} > 5$	Extremely contaminated

2.5. Principal Component Analysis (PCA)

The principal component analysis (PCA) adopted in this study is a dimensionality reduction method that applies a linear transformation to numerous interrelated original variables. It extracts fewer essential variables that are uncorrelated with each other. Using fewer representative factors, PCA can explain the primary information of multiple variables and infer relevant pollution sources, enabling the more accurate qualitative identification of pollution sources [34].

2.6. Positive Matrix Factorization (PMF)

The positive matrix factorization (PMF) employed in this study is a receptor model recommended by the US EPA and first introduced by Paatero and Tapper in 1994 [35]. This model identifies pollution sources using sample composition or fingerprints, applying non-negative constraints during the factor matrix decomposition process to achieve a quantitative analysis of pollution sources. In the PMF model, the concentration data matrix of receptor samples (X) is decomposed into a factor score matrix (G), a factor loading matrix (F), and a residual matrix (E). The matrix form is expressed as follows (Equations (2) and (3)):

$$X = GF + E \tag{2}$$

$$x_{ij} = \sum_{k=1}^{p} g_{ik} f_{kj} + e_{ij} \tag{3}$$

Here, x_{ij} represents the concentration of compound j in sample i (mg· kg^{-1} d.w.); g_{ik} denotes the contribution of source k to sample i; f_{kj} is the concentration of compound j in source k (mg· kg^{-1} d.w.); p represents the total number of identified pollution sources in the model; and e_{ij} is the residual for each sample and compound. The results of the PMF model are determined by the objective function (Equation (4)).

$$Q(E) = \sum_{i=1}^{n} \sum_{j=1}^{m} \left(\frac{e_{ij}}{u_{ij}} \right)^2 \tag{4}$$

Here, u_{ij} represents the uncertainty of each compound in each sample, which is dimensionless, and the method for calculating uncertainty refers to the literature [36].

When the concentration of potentially toxic elements is below the detection limit of the corresponding method, the parameters are suggested as follows (Equation (5)):

$$x_{ij} = \frac{d_{ij}}{2},\ u_{ij} = \frac{5}{6} d_{ij} \tag{5}$$

In reverse, the parameters can be set as follows. The value x_{ij} can be seen as the PTE concentration detected. The uncertainty factor will be chosen depending on Equation (6) or (7).

$$x_{ij} \leq 3d_{ij}, \quad u_{ij} = \frac{d_{ij}}{3} + 0.2 \times c_{ij} \tag{6}$$

$$x_{ij} > 3d_{ij}, \quad u_{ij} = \frac{d_{ij}}{3} + 0.1 \times c_{ij} \tag{7}$$

Here, c_{ij} is the measured concentration of the sample, mg· kg^{-1} d.w.; values of x_{ij} are the input concentration during the model operation process, mg· kg^{-1} d.w.; d_{ij} is the detection limit of the method, mg· kg^{-1} d.w.; and u_{ij} is the uncertainty factor, dimensionless.

2.7. Data Analysis

The concentrations of each PTE were statistically analyzed using Microsoft Excel 2016. Subsequently, all test data were subjected to a normality test using the Kolmogorov–

Smirnov test within SPSS 18.0 software, and the results were visualized using OriginPro 2024. The data are considered to have a normal distribution if the *p*-value is more significant than 0.05. For source apportionment analysis, the PCA module of SPSS 18.0 is employed in conjunction with PMF 5.0, downloaded from the US EPA website [37].

3. Results and Analysis

3.1. Descriptive Statistics of PTEs in Soil

The profiles of PTE levels in agricultural soil were surveyed (Table 2). The descending order of mean value in the soil is as follows: Mn, Zn, Cu, Pb, Cr, As, Ni, Sb, Mo, and Cd. It is noteworthy that the average concentrations of Mo and Pb are marginally elevated above the corresponding soil background values in Jiangxi Province. Also, the average of Cu and Cd are substantially higher, with 3.76 and 5.40 times the background levels, respectively. Of particular concern, over 80% of the surveyed sites had Cu and Cd concentrations surpassing the background values. Furthermore, compared to the risk screening values for soil contamination on agricultural land, as stipulated by the standard GB 15618-2018 [38], this indicates that the average of Cu and Cd are 1.53 and 1.97 times, respectively.

The coefficient of variation (CV) is defined as the degree of spatial variability in PTE concentrations [39], and it is sorted as low (CV $\leq$ 0.30), moderate (0.30 < CV $\leq$ 0.50), high (0.50 < CV $\leq$ 1.00), or extremely high (CV > 1.00). The CV values for PTEs in this study range from 0.51 to 2.56, with an average of 1.33 (Table 2), indicating an extremely high level of variation in the study area, which might be affected by external factors.

Table 2. Statistical analysis of PTE concentrations in soil.

	Mean of All Soil (1) (mg·kg^{-1} d.w.)	Minimum (1) (mg·kg^{-1} d.w.)	Maximum (1) (mg·kg^{-1} d.w.)	Variation Coefficient	Soil Background Value in Jiangxi Province (2) (mg·kg^{-1})	Risk Screening Value (3) (mg·kg^{-1})	Average of the *Igeo*
Mn	137.88	14.63	1861.00	1.31	328.00	-	−2.43
Zn	89.62	25.81	2717.60	2.56	69.40	200.00	−0.77
Cu	76.30	10.23	739.05	1.39	20.30	50.00	0.73
Pb	35.56	7.47	990.83	2.48	32.30	80.00	−1.06
V	20.58	4.59	96.76	0.65	95.80	-	−3.02
Cr	15.47	3.49	57.14	0.51	45.90	250.00	−2.31
As	8.57	1.13	116.05	1.60	14.90	30.00	−1.91
Ni	8.42	2.13	28.55	0.57	18.90	60.00	−1.93
Co	3.21	0.66	11.36	0.65	11.50	-	−2.70
Sb	0.90	0.02	9.92	1.26	1.15	-	−1.73
Mo	0.66	0.16	4.19	0.98	0.50	-	−0.56
Cd	0.59	0.03	10.85	2.00	0.11	0.30	0.84

Notes: (1) The concentrations of the elements analyzed are reported on a dry weight basis (mg· kg^{-1} d.w.). (2) Soil background value of layer A in Jiangxi Province (background values of soil elements in China, China National Environmental Monitoring Centre, 1990). (3) Risk screening values for soil contamination of agricultural land, pH $\leq$ 5.5 (Soil Environmental Quality Risk Control Standard for Soil Contamination of Agricultural Land, GB15618-2018, Ministry of Ecology and Environment of China, 2018).

3.2. Assessment of Soil PTs Contamination

To further investigate the degree of PTE pollution in agricultural soils, the Index of Geo-accumulation was applied. The regional *Igeo* mean is ranked from lowest to highest (Figure 2a), listed as V < Co < Mn < Cr < Ni < As < Sb < Pb < Zn < Mo < Cu < Cd. The pollution status of Cu and Cd in the soil is particularly concerning, with their respective *Igeo* mean values of 0.73 and 0.84, both falling within the class 1 zone (Table 1), which is a scope of moderate contamination. In contrast, all other PTEs fall in the zone of class 0, with no pollution. As far as how PTEs in soil fall into the scope of contamination is concerned, the *Igeo* indicates that Cu, Cd, Mo, and Sb are 74.32%, 70.95%, 25.00%, and 11.49%, respectively, whereas Mn, Ni, As, Pb, and Zn are below 10%. No contamination of V, Co, or Cr is found (Figure 2b). After calculating the *Igeo* value, it emphasizes the necessity to pay more attention to the risk that Cu and Cd may pose.

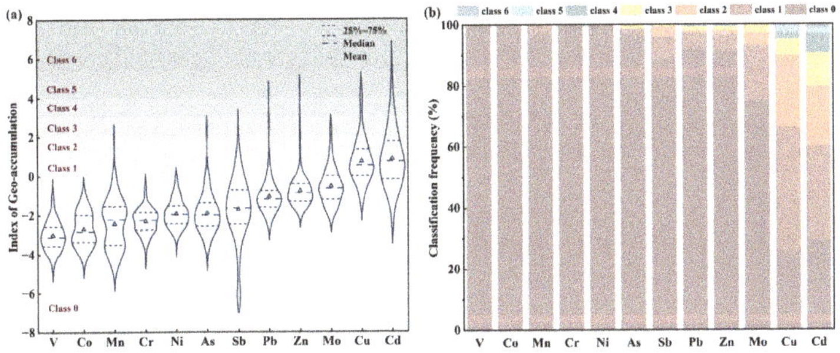

Figure 2. Index of Geo-accumulation in the study area.

3.3. *Source Apportionment of Soil PTEs*

3.3.1. Homology Analysis of Soil PTEs

The theory of elemental geochemistry puts up a view that the correlation among elements can reflect the similarities in their sources, and accordingly, the homogeneity analysis of PTEs can provide a basis for identifying the sources of PTEs [40–42], which is performed using correlation statistics and clustering analysis [40,43]. The content of soil PTEs in the study area was found to be normally distributed, determined by R-type clustering and Spearman correlation analysis.

Figure 3a shows that PTEs are clustered into three types: Cluster I includes V, Cr, Co, and Ni; Cluster II includes Mn, Zn, Pb, and Cd; and Cluster III includes Cu, Mo, Sb, and As. The inter-cluster elements have significant autocorrelation, indicating that these PTEs may have homologous sources [44]. In conjunction with the correlation heat map (Figure 3b), it can be inferred that V, Cr, Co, and Ni in Cluster I may have homologous sources. Similarly, Mn, Zn, Pb, and Cd in Cluster II and Cu, Mo, Sb, and As in Cluster III are speculated as having homologous sources. A comparative analysis across the three clusters shows poor correlation among the elements in different clusters, suggesting a lower homology among these clusters. As a result, it can be concluded that there are at least three potential sources of PTEs in agricultural soils.

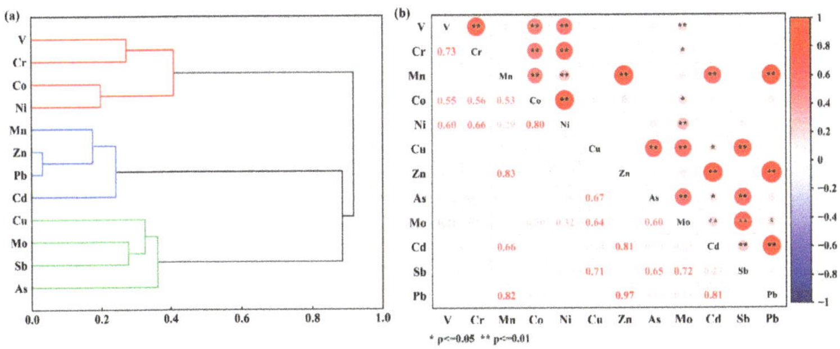

Figure 3. Homology analysis of potentially toxic elements in soil. (**a**,**b**) R-type cluster map and Spearman correlation heat map, respectively.

3.3.2. PCA Analysis of Soil PTEs

PCA is further utilized to delineate the sources of PTEs in the soil. The Kaiser–Meyer–Olkin test gave a value of 0.76, a value suitable for factor analysis, and the Bartlett sphericity examination passed. Following varimax rotation to maximize variance, three principal components were identified, adhering to the criterion of eigenvalues exceeding 1 in PCA. These components cumulatively account for 80.34% of the total variance, suggesting that the majority of PTE sources can be attributed to these three principal components. The three factors after rotation are shown (Table 3), and the loadings on each factor are displayed (Figure 4).

Table 3. Rotating loadings of potentially toxic elements on principal components.

Elements	Component		
	PC1	PC2	PC3
V	−0.06	**0.84**	0.02
Cr	−0.07	**0.85**	0.03
Mn	**0.89**	0.27	−0.06
Co	0.23	**0.85**	−0.01
Ni	0.07	**0.88**	0.14
Cu	−0.02	−0.13	**0.89**
Zn	**0.97**	0.01	0.03
As	0.08	0.09	**0.83**
Mo	0.13	0.24	**0.83**
Cd	**0.87**	−0.08	0.21
Sb	0.06	−0.01	**0.89**
Pb	**0.97**	−0.01	0.08
Eigenvalues	4.00	2.88	2.76
Variance contribution ratio %	29.29	25.70	25.36
Cumulative contribution ratio %	29.29	54.98	80.34

Note: (1) The rotation method used varimax with Kaiser Normalization, and the rotation converged in 4 iterations; (2) the bold numbers represent the higher values of the load in each principal component.

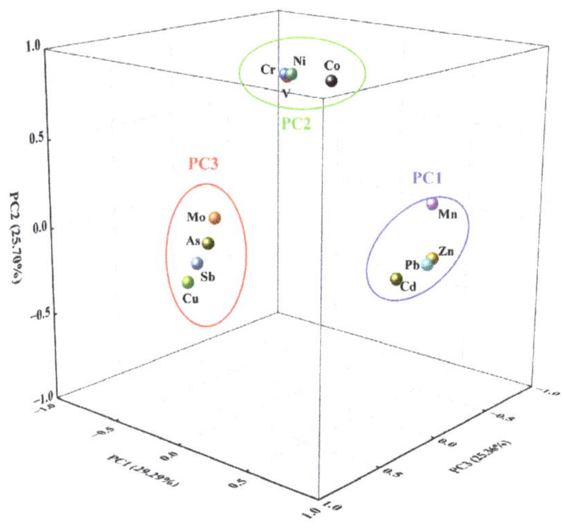

Figure 4. PCA loading contribution of potentially toxic elements.

The first principal component (PC1) explains 29.29% of the total variance, with high loadings (>0.85) observed for Mn, Zn, Cd, and Pb, which indicates that these elements likely share a homologous source or similar transport pathways, as evidenced by their significant correlations (Figure 3). Based on the Mineral Resource Master Plan of Yingtan City (2008–2015), it is confirmed that the area boasts a significant abundance of lead–zinc mineral resources. According to previous research, PTEs such as Pb, Zn, and Cd are released into the environment during the mining process and subsequently enter the soil through weathering, erosion, and other environmental processes [45–47]. Also, a large amount of wastewater and slag is discharged during the ore dressing process, which contains high concentrations of PTEs [46]. For example, in a typical lead–zinc mining area in Guizhou Province, the average concentrations of Pb and Zn in the ore dressing wastewater reach 10.5 mg/L and 25.3 mg/L, respectively, far exceeding the national discharge standards (GB 8978-1996) [48,49]. If these wastewaters are discharged directly without treatment, they can cause severe pollution to the soil and groundwater. The research area has a developed mining industry, which might be one of the pollution source contributors. Moreover, Cd has been identified as an element associated with agricultural practices, and the excessive use of phosphate fertilizers can result in Cd accumulation in soil [50,51]. Similar fertilizers were applied in the investigation areas; therefore, PC1 is assigned as a composite source encompassing both mining and agricultural activities.

The second principal component (PC2) explains 25.70% of the total variance, in which strongly positive correlations of V, Cr, Co, and Ni were explored, and the CV values of the four elements were all relatively weak (0.51~0.65); meanwhile, their average contents did not exceed the soil background values in Jiangxi Province. All the data suggest that external activities have minimal impact on their distribution. Concerning the source of V, Cr, Co, and Ni, the previous report indicated that natural sources originating from soil matrix matrices and weathering processes may be the main input [30,32,52,53]. Combined with the level of background PTEs in Jiangxi Province, it is speculated that PC2 may be a natural source.

The third principal component (PC3) explains 25.36% of the total variance, with Cu, As, Mo, and Sb being the main loading elements. Among them, Cu pollution is the most severe, with an average content exceeding 3.76 times the soil background value in Jiangxi Province. According to Zhou [54], the Guixi Smelter did not plan and control the discharge of waste residue, wastewater, and exhaust gas generated during the smelting process in the early stages of operation. After more than 30 years of accumulation, it has caused pollution to the surrounding environment, with the main pollutants including Cu. The investigation conducted by Wu et al. [55] and Xu et al. [56] also confirms this statement. In addition, substantial acid waste containing As, Mo, and Sb was generated from copper smelting and released into the environment by slag, wastewater, and flue gas [57,58]. These PTEs enter the soil via atmospheric deposition, waste residue leakage, sewage irrigation, and other pathways, resulting in enrichment in the soil. Thus, the main source of elements of PC3 can be identified as copper smelting activities.

3.3.3. PMF Analysis of Soil PTEs

To further elucidate the origins and contributions of PTEs in agricultural soils, PMF was employed to determine the sources. The PTE concentration data and their corresponding uncertainty parameters were introduced into the PMF model. Based on the results from PCA, the parameters of three primary sources were set, but an unsatisfactory fitting was obtained. To improve the fitting degree, more source numbers were selected to reflect the multiple sources of soil PTEs in the research area, and the Robust model was adopted. The number of runs was set as 20, and the mode of "random starting seed number" was optioned in PMF, and then parameters of five source factors were identified. The fitting results between the real and predicted values are presented in Table 4. All PTEs exhibit high correlation coefficients (R^2) above 0.50, indicating that the PMF simulation yields well-fitting results in response to the sources.

Table 4. Fitting results of measured value and simulated predicted value of potentially toxic element content.

Elements	R^2	Intercept	Slope
V	0.68	6.92	0.60
Cr	0.85	1.82	0.85
Mn	0.64	58.86	0.47
Co	0.88	0.62	0.72
Ni	0.68	2.88	0.62
Cu	0.86	12.21	0.76
Zn	0.76	47.95	0.19
As	0.63	3.91	0.38
Mo	0.70	0.17	0.63
Cd	0.87	0.02	0.89
Sb	0.80	0.04	0.80
Pb	0.59	18.45	0.14

In PMF, the loading factors for each potentially toxic element can help to apportion the sources (Figure 5), and it can be observed that PM1 is more loaded on Mn. However, the average concentration of Mn does not exceed the background values for Jiangxi Province, and only a few sample sites are found in a state of pollution. Regarding the sources, natural sources are the first focus. Mn, a constituent of the Earth's crust, is naturally abundant in soils. Yingtan, a city with abundant manganese ore resources, is one of the main production areas of manganese ore in China. According to official statistical data, the manganese ore production in Yingtan has remained at more than 300,000 tons in the past few years. Most of these mines are open mines, and Mn would be released into the environment along with slag disposal and atmospheric sedimentation, which explains the increasing Mn soil in the local area. Considering the fact of mining activities, it is inferred that PM1 is mainly influenced by natural origin and mining activities.

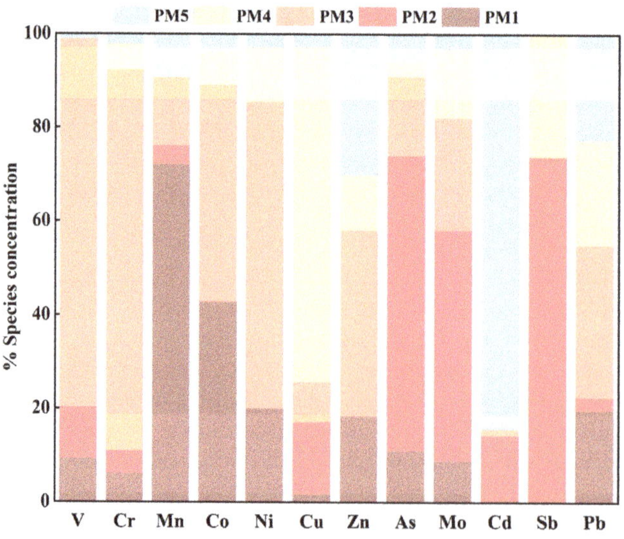

Figure 5. Factor fingerprints of potentially toxic elements.

The main loading factors of PM2 are Sb, As, and Mo, which overlap with the composition of PC3 in PCA. These elements are tightly associated with the by-products of copper smelting. Thus, PM2 is apportioned as an industrial source relating to copper smelting activities.

In PM3, the contribution ratios are dominated by V, Cr, Ni, Co, Zn, and Pb, with loading factor ratios of 78.67%, 81.14%, 65.41%, 46.12%, 39.58%, and 32.57%, respectively. It is worth noting that there is no correlation between the group of Zn and Pb and another group of V, Cr, Ni, and Co, which mainly come from soil parent materials and are less affected by human activities [30,32,52,53]. The study area was found to be rich in lead–zinc minerals and mining activities. In this study, the average contents of V, Cr, Ni, and Co are lower than the corresponding soil background values in Jiangxi Province (Table 2), with a low CV value, which means these PTEs come from natural sources, while soil Zn and Pb are contributed by anthropology activities. Therefore, PM3 can be characterized as a mixed source of natural origin and mining activities.

PM4 was mainly characterized by loadings from Cu. In this study, over 80% of the survey sites showed Cu concentrations exceeding the background value, with an average exceedance multiple of 3.76, which indicates that Cu has been related to severe external input. The copper smelting activities mentioned may have a role in contributing to Cu in soil. As far as other input ways, it cannot be ignored that Cu, as an inherent component of additives in livestock feed, will be transferred to the soil via animal manure [59]. In addition, using fertilizers and pesticides can lead to the accumulation of Cu in the soil [60]. So, PM4 is set as a mixed source of agriculture and copper smelting.

PM5 has a relatively high contribution of 84.37% to Cd. It was often regarded as a hallmark element in agricultural activities [50,51]. In general, some research found that phosphate ores contain 5–100 mg/kg of Cd, and most or all of the Cd enters the fertilizer [51]. Excessive application of phosphate fertilizer can cause the accumulation of Cd in the soil. Hence, PM5 is considered an agricultural source.

3.3.4. Integrating PCA-PMF with Field Survey

The source apportionment and counterpart contribution ratios obtained from both PCA and PMF are shown (Figure 6). PCA has accounted for the sources of PTEs in soil by using three factors, while PMF, based on three factors, has yielded suboptimal model fitting because of the complexity of input sources. Due to a complex environmental background, the pathways through which PTEs enter the soil are diverse, making it challenging to define them with a single source. The PMF model provides an advantage to overcome the shortage by optimizing the parameters of input sources and uncertainties. In terms of the source apportionment results, it can be observed that despite the use of five factors by PMF to describe the sources of PTEs, which still include the PCA analysis, the sources are mining activity, natural, smelting, and agricultural activity. This integration of PCA-PMF can be cross-validated for allocating pollution sources, which is more suitable for dealing with the complexity of PTE sources in soil.

The escalating pollution of agricultural soils with Cd and Cu not only impacts crop growth but also poses a prolonged risk to human health through the food chain [61,62]. This study underscores that the contamination of Cu and Cd in the study area's soil warrants significant attention (Figure 2). For the specific area, an extra field survey combined with PCA-PMF can help apportion sources. For example, ZF Town, a green farmland area far away from the industrial zones, has found that Cd rice exceeded the standard. It is important to identify the source of PTEs. Information on historical land use in ZF is inquired, and the soil can be classified into two categories: soil with mixed fertilizer application (SF) and soil without mixed fertilizer application (SNF). The technology of

PCA-PMF was employed in the specific green agricultural area, and the results are shown in Figure S2 and Table S4. The results show that Cd and Cu were alien pollution elements in farmland soils, and there was a significant correlation between them (Figure S1). In the previous analysis, it has been pointed out that soil Cd and Cu are closely related to the application of fertilizers. However, an interesting phenomenon was found in Table S1. The average of Cd in SNF soil is 0.06 mg· kg^{-1} d.w., which is considerably lower than the 0.18 mg· kg^{-1} d.w. observed in SF soil and the soil background value of 0.11 mg·kg^{-1} in Jiangxi Province. In contrast, the mean Cu concentration in SNF soil is 30.21 mg· kg^{-1} d.w., significantly lower than the 39.09 mg· kg^{-1} d.w. in SF soil, but it exceeds the background value of 20.3 mg·kg^{-1}. This result indicates that fertilization is likely the predominant source of Cd, whereas Cu, although also present in higher concentrations in SF soil, may have additional anthropogenic sources contributing to its levels in the soil. For example, the long-term application of pig manure can lead to Cu accumulation in farmland soil [63,64]. This methodology of PCA-PMF merged with field survey will enable a more comprehensive understanding of the origins of Cd and Cu, particularly in the agricultural area.

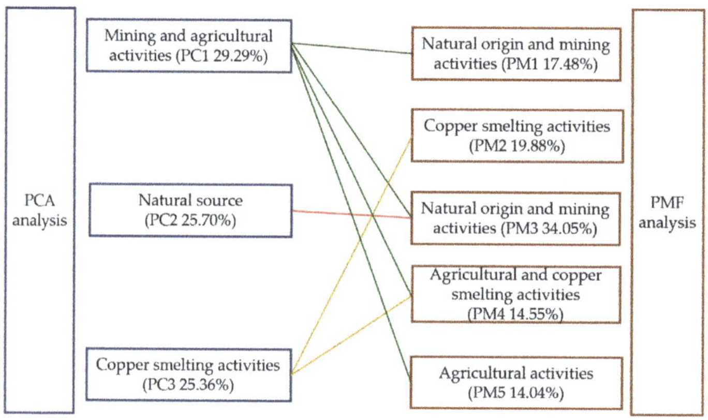

Figure 6. A comparison of source apportionment by PCA and PMF.

4. Discussion

Cd and Cu contamination in farmland soil poses significant risks to human and environmental health. Cd, a toxic metal, enters the human body through the food chain, causing kidney damage, bone deformities (e.g., Itai-Itai disease), and increased cancer risk with long-term exposure [65]. It also impairs soil biota and plant growth, reducing biodiversity and ecosystem services [66]. Cu, essential in small amounts, becomes harmful at high levels, causing gastrointestinal, liver, and kidney issues in humans and disrupting the nervous system. In soil, elevated Cu levels inhibit microbial activity that is crucial for nutrient cycling and plant growth, leading to reduced crop yields and quality [67].

Given the complex pollution distribution in the study area, a multi-method approach is recommended for highly polluted small-area farmland patches. Deep plowing combined with guest-soil methods can dilute contaminants. Acidic modifiers like ferrous sulfate can adjust soil pH to precipitate Cd and Cu without harming the acidic environment. For bioremediation, local acid-tolerant hyperaccumulators such as Rumex species, which excel in Cd enrichment, can be planted in treated soils to boost phytoremediation efficiency. Additionally, acid-tolerant microorganisms like Acid thiobacillus can be introduced, which can transform Cd and Cu to enhance the removal of PTEs when cooperating with plants.

This highlights the need for government departments to enhance the environmental supervision of the copper-smelting industry, drive smelters to upgrade pollution-control facilities, ensure standard waste discharge, and curb PTE migration to farmland. They should strictly enforce ecological protection for mining, regulate mining procedures, improve drainage systems, and prevent PTE wastewater from entering farmland. Ecological restoration, like vegetation cover on abandoned mines, should be implemented to block PTE spread. In agriculture, scientific fertilization and pesticide use should be promoted. Organic fertilizer use should be encouraged to replace some chemical fertilizers, PTE input from fertilizer impurities should be reduced, and farmers should be guided on using low-toxicity, low-residue pesticides with minimal impact on soil PTEs, building a PTE source-control system.

Additionally, current research has limitations in scale, and relying solely on PCA and PMF may not fully capture the complexity of PTs sources, especially considering the influence of various environmental factors and the dynamic nature of soil interactions. It is recommended that introducing isotope methods could provide a more robust framework for source identification by offering geochemical fingerprints that are less affected by external conditions and more closely related to the compositional features of the source region.

5. Conclusions

This study suggests that the input of PTEs into agricultural soil may be attributed to mining activity, natural sources, copper smelting, and agricultural practices. Among these elements, Cu and Cd were significantly enriched in agricultural soils and identified as the most critical pollutants in the region. The comparative analysis using PCA and PMF demonstrated a high degree of congruence in identifying pollution sources, providing mutual validation for the large-scale region with complex sources. PCA excels in determining the primary number of predominant pollution sources, while PMF is more precise in calculating their contribution rates. For specific sites, the integration of PCA-PMF with field investigations effectively pinpoints the sources and has proven to be a robust approach in this study.

Supplementary Materials: The following supporting information can be downloaded at: https://www.mdpi.com/article/10.3390/toxics13040267/s1, Table S1: Statistical analysis of PTEs concentrations in the typical green agricultural soil; Table S2: Rotating loadings of PTEs on principal components; Table S3: Fitting results of measured value and simulated predicted value of PTEs content based on PMF; Table S4: Sources spectra and source contribution of PTEs based on PMF; Figure S1: The homology analysis of PTEs in the typical green agricultural soil; Figure S2: PCA loading contribution of the PTEs.

Author Contributions: Conceptualization, H.W.; data curation, R.H.; writing—original draft preparation, S.C.; writing—review and editing, H.W. All authors have read and agreed to the published version of the manuscript.

Funding: This research was funded by the National Key Research and Development Program of China (2021YFC1809104).

Institutional Review Board Statement: Not applicable.

Informed Consent Statement: Not applicable.

Data Availability Statement: The raw data supporting the conclusions of this article will be made available by the authors upon request.

Conflicts of Interest: The authors declare no conflicts of interest.

References

1. Chen, S.B.; Wang, M.; Li, S.S.; Zheng, H.; Lei, X.Q.; Sun, X.Y.; Wang, L.F. Current status of and discussion on farmland heavy metal pollution prevention in China. *Earth Sci. Front.* **2019**, *26*, 35–41. [CrossRef]

2. Ministry of Environmental Protection; Ministry of Land and Resources. National Soil Pollution Status Survey Bulletin. Available online: https://www.gov.cn/foot/site1/20140417/782bcb88840814ba158d01.pdf (accessed on 15 September 2023).
3. Zhao, J.X.; Yin, P.C.; Yue, R.; Wang, M.P.; Shi, R. Research progress of status, source, restoration technique of heavy metals pollution in cropland of China. *J. Anhui Agric. Sci.* **2018**, *46*, 19–21+26. [CrossRef]
4. Wang, Y.; Zhang, X.L.; Wang, R. Soil pollution survey and management in Taiwan of China. *Soils* **2018**, *50*, 7–15. [CrossRef]
5. Hu, Y.A.; He, K.L.; Sun, Z.H.; Chen, G.; Zheng, H.F. Quantitative source apportionment of heavy metal(loid)s in the agricultural soils of an industrializing region and associated model uncertainty. *J. Hazard. Mater.* **2020**, *391*, 122244. [CrossRef]
6. Yu, F.R.; Ji, Y.K.; Wu, L.; Zhang, J.N.; Fu, Y.; Liu, S.T.; Ma, J.Q. Source-oriented comprehensive assessment framework for identifying priority heavy metals in agricultural soils. *J. Environ. Chem. Eng.* **2025**, *13*, 115579. [CrossRef]
7. Yang, L.; Bai, Z.X.; Bo, W.H.; Lin, J.; Yang, J.J.; Chen, T. Analysis and evaluation of heavy metal pollution in farmland soil in China: A Meta-analysis. *Environ. Sci.* **2024**, *45*, 2913–2925. [CrossRef]
8. Fei, X.F.; Lou, Z.H.; Xiao, R.; Ren, Z.Q.; Lv, X.N. Source analysis and source-oriented risk assessment of heavy metal pollution in agricultural soils of different cultivated land qualities. *J. Clean. Prod.* **2022**, *341*, 130942.
9. NI, Z.Y.; Xie, G.X.; Zhang, M.K. Effects of Acidification on Bioavailability of Cadmium and Lead in Cultivated Land Soil and Their Accumulation in Agricultural Products. *Acta Agric. Jiangxi* **2017**, *29*, 52–56.
10. Wang, Q.; Pang, Y.; Xu, Y.; Yuan, Y.; Yin, D.; Hu, M.; Xu, L.; Liu, T.; Sun, W.; Yu, H.Y. Controlling factors of heavy metal(loid) accumulation in rice: Main and interactive effects. *Environ. Sci. Pollut. Res.* **2024**, *31*, 42357–42371.
11. Liu, Q.; Lu, W.; Bai, C.; Xu, C.; Ye, M.; Zhu, Y.; Yao, L. Cadmium, arsenic, and mineral nutrients in rice and potential risks for human health in South China. *Environ. Sci. Pollut. Res.* **2023**, *30*, 76842–76852. [CrossRef]
12. Zhou, L. Research on the Content and Distribution of Heavy Metals in the Soil of a Vegetable Base Site. Master's Thesis, Nanchang University, Nanchang, China, 2016.
13. Lv, Y.; Xie, L.; Zhu, W.; Zhou, Y.; Sun, H. Early warning research on the probability of soil heavy metal risk in county agricultural land based on environmental capacity. *Resour. Environ. Yangtze River Basin* **2020**, *29*, 253–264.
14. Wan, Y.; Liu, J.; Henríquez-Hernández, Z.Z.W.L.A. Heavy Metals in Agricultural Soils: Sources, Influencing Factors, and Remediation Strategies. *Toxics* **2024**, *12*, 63. [CrossRef] [PubMed]
15. Wang, H.; Wu, J.; Tian, Z. Study on atmospheric heavy metal deposition by environmental tracers surrounding copper smelting. *Air Qual. Atmos. Health* **2023**, *16*, 1479–1487. [CrossRef]
16. Huang, Y.; Chen, G.; Huang, Y.; Liu, B.; Xiong, L.; Xiaoqing, O.U.; Zhao, X.; Zhou, Y.; Pang, M. Characteristics and Risk Assessment of Soil Heavy Metal Contents in Facility Agriculture of Guangxi. *Asian Agric. Res.* **2021**, *13*, 5.
17. Huang, Y.; Chen, Q.; Deng, M.; Japenga, J.; Li, T.; Yang, X.; He, Z. Heavy metal pollution and health risk assessment of agricultural soils in a typical peri-urban area in southeast China. *J. Environ. Manag.* **2018**, *207*, 159–168. [CrossRef]
18. Zhang, X. Sources Analysis of Heavy Metal Contamination in Soil and Rice in an Industrial Area in Hunan Province, China. Master's Thesis, Chinese Academy of Agricultural Sciences, Beijing, China, 2019.
19. Wei, Y.H.; Li, G.C.; Wang, Y.H.; Zhang, Q.; Li, B.; Wang, S.C.; Cui, J.H.; Zhang, H.Z.; Zhang, Q. Investigating factors influencing the PMF model:A case study of source apportionment of heavy metals in farmland soils near a lead-zinc ore. *J. Agro-Environ. Sci.* **2018**, *37*, 2549–2559.
20. Diakite, M.L.; Hu, Y.; Cheng, H. Source apportionment based on the comparative approach of two receptor models in a large-scale region in China. *Environ. Sci. Pollut. Res.* **2021**, *28*, 56696–56710. [CrossRef]
21. Zhang, X.; Wei, S.; Sun, Q.; Wadood, S.A.; Guo, B. Source identification and spatial distribution of arsenic and heavy metals in agricultural soil around Hunan industrial estate by positive matrix factorization model, principle components analysis and geo statistical analysis. *Ecotoxicol. Environ. Saf.* **2018**, *159*, 354–362. [CrossRef]
22. Jiang, F.; Ren, B.; Hursthouse, A.S.; Zhou, Y. Trace Metal Pollution in Topsoil Surrounding the Xiangtan Manganese Mine Area (South-Central China): Source Identification, Spatial Distribution and Assessment of Potential Ecological Risks. *Int. J. Environ. Res. Public Health* **2018**, *15*, 2412. [CrossRef]
23. Jiang, Y.; Chao, S.; Liu, J.; Yang, Y.; Chen, Y.; Zhang, A.; Cao, H. Source apportionment and health risk assessment of heavy metals in soil for a township in Jiangsu Province, China. *Chemosphere* **2017**, *168*, 1658–1668. [CrossRef]
24. Đorđević, D.; Petrović, S.; Relić, D.; Mihajlidi-Zelić, A. Applying receptor models Unmix and PMF on real data set of elements in PM for sources evaluation of the sea coastal side region (Southeast Adriatic Sea). *Atmos. Meas. Tech. Discuss.* **2013**, *2013*, 4941–4969. [CrossRef]
25. Wang, C.; Zhao, Y.P.; Xie, M.J. Characteristics of lead enrichment in the soil from a typical peri-urban agricultural area of the southern Jiangsu and source appointment based on the PCA-PMF method. *China Environ. Sci.* **2021**, *41*, 279–287. [CrossRef]
26. Yang, Z.; Yu, J.; Yang, K.; Zhang, Q.; Chen, Y.; Qiao, S. Source Apportionment and Risk Assessment of Potentially Toxic Elements Based on PCA and PMF Model in Black Soil Area of Hailun City, Northeast China. *Toxics* **2024**, *12*, 683. [CrossRef]
27. *GB 2762-2012*; National Food Safety Standard—Limits of Contaminants in Food. National Health Commission of the People's Republic of China: Beijing, China, 2012.

28. *HJ 803-2016*; Soil and Sediment-Determination of Aqua Regia Extracts of 12 Metal Elements-Inductively Coupled Plasma Mass Spectrometry. Ministry of Ecology and Environment of the Reople's Republic of China: Beijing, China, 2016.
29. Müller, G. Index of Geoaccumulation in Sediments of the Rhine River. *GeoJournal* **1969**, *2*, 109–118.
30. Luo, H.P.; Wang, Q.Z.; Guan, Q.Y.; Ma, Y.R.; Zhang, J. Heavy metal pollution levels, source apportionment and risk assessment in dust storms in key cities in Northwest China. *J. Hazard. Mater.* **2021**, *422*, 126878.
31. Li, Z.Y.; Ma, Z.W.; van der Kuijp, T.J.; Yuan, Z.W.; Huang, L. A review of soil heavy metal pollution from mines in China: Pollution and health risk assessment. *Sci. Total Environ.* **2014**, *468–469*, 843–853. [CrossRef]
32. Pan, H.Y.; Lu, X.W.; Lei, K. A comprehensive analysis of heavy metals in urban road dust of Xi'an, China: Contamination, source apportionment and spatial distribution. *Sci. Total Environ.* **2017**, *609*, 1361–1369. [CrossRef]
33. Anaman, R.; Peng, C.; Jiang, Z.C.; Liu, X.; Zhou, Z.R.; Guo, Z.H.; Xiao, X.Y. Identifying sources and transport routes of heavy metals in soil with different land uses around a smelting site by GIS based PCA and PMF. *Sci. Total Environ.* **2022**, *823*, 153759. [CrossRef]
34. Kuang, H.F.; Hu, C.H.; Wu, G.L.; Chen, M. Combination of PCA and PMF to apportion the sources of heavy metals in surface sediments from Lake Poyang during the wet season. *J. Lake Sci.* **2020**, *32*, 964–976.
35. Paatero, P.; Tapper, U. Positive matrix factorization: A non-negative factor model with optimal utilization of error estimates of data values. *Environmetrics* **1994**, *5*, 111–126.
36. Zhang, J.Z.; Zhou, X.H.; Wang, Z.; Yang, L.X.; Wang, J.; Wang, W.X. Trace elements in PM2.5 in Shandong Province: Source identification and health risk assessment. *Sci. Total Environ.* **2018**, *621*, 558–577.
37. U.S. Environmental Protection Agency. Available online: https://www.epa.gov/air-research/positive-matrix-factorization-model-environmental-data-analyses (accessed on 15 September 2023).
38. *GB 15618-2018*; Soil Environmental Quality Risk Control Standard for Soil Contamination of Agricultural Land. Ministry of Ecology and Environment of the Reople's Republic of China: Beijing, China, 2018.
39. Wang, H.Z.; Liu, D.Y.; Lv, Y.F.; Wang, W.; Wu, Q.; Huang, L.; Zhu, L. Ecological and health risk assessments of polycyclic aromatic hydrocarbons (PAHs) in soils around a petroleum refining plant in China: A quantitative method based on the improved hybrid model. *J. Hazard. Mater.* **2024**, *461*, 132476. [CrossRef]
40. Xie, Z.J.; Zhao, R.B.; Wang, Y.; Wang, H.; Li, C.H.; Yu, Z.L.; Ye, C. Ecological risk assessment and pollution characteristic analysis of heavy metals in sediments of Xingkai Lake. *J. Environ. Eng. Technol.* **2023**, *13*, 1987–1996.
41. Lv, Y.J.; Wang, Q.Y.; Sun, X.M.; Zhang, Z.W.; Zhang, Y.M.; Gao, Y.X. Pollution characteristics and source identification of heavy metals in farmland soils around a tailing pond in Zhejiang Province. *J. Environ. Eng. Technol.* **2023**, *13*, 1464–1475.
42. Mi, X.J.; Ren, W.; Luo, Q.; Ma, J. Evaluation and their sources of heavy metals in uncultivated saline-alkaline soil in the Junggar Basin, Xinjiang. *Arid Zone Res.* **2019**, *36*, 824–834. [CrossRef]
43. Shuai, W.C.; Liu, W.Q.; Ma, L.Y.; Ming, C.H.; Chen, J.B.; Yuan, W. Analysis of heavy metal sources and ecological health risk assessment in typical non-ferrous metal smelting sites. *Earth Environ.* **2024**, *52*, 756–770. [CrossRef]
44. Li, X.X.; Liu, H.Y.; Meng, W.; Liu, N.T.; Wu, P. Accumulation and source apportionment of heavy metal(loid)s in agricultural soils based on GIS, SOM and PMF: A case study in superposition areas of geochemical anomalies and zinc smelting, Southwest China. *Process Saf. Environ. Prot.* **2022**, *159*, 964–977.
45. Liu, A.; Yu, C.L.; Wang, L.P.; Song, J.J.; Sun, L.W.; Qian, J.; Sun, M.H. Heavy metal sources and ecological risk assessment of typical lead-zinc mining areas in Hebei Province. *Bull. Geol. Sci. Technol.* **2024**, *43*, 307–317. [CrossRef]
46. Chen, Y. Research progress on soil heavy metal pollution and ecological restoration technology in lead-zinc mining areas. *China Strateg. Emerg. Ind.* **2024**, *32*, 122–124.
47. Liu, N.; Tang, Y.Y.; Chen, M.; Pan, Y.X. Source apportionment of soil heavy metals in lead-zinc area based on APCS-MLR and PMF. *China Environ. Sci.* **2023**, *43*, 1267–1276.
48. Gu, H.; Zhao, T.; Gao, Y.; Sun, R.G. Pollution characteristics and source analysis of heavy metals in soils of a typical lead-zinc mining Area in Guizhou Province. *Earth Environ.* **2022**, *50*, 506–515. [CrossRef]
49. *B 8978-1996*; Integrated Wastewater Discharge Standard. Ministry of Ecology and Environment of the Reople's Republic of China: Beijing, China, 1996.
50. Yang, Y.; Zhou, J.; Guo, T.; Ding, H.; Liu, X. Source analysis of heavy metal pollution in farmland soil in a mining area based on a small watershed scale. *J. Agro-Environ. Sci.* **2023**, *42*, 1956–1963.
51. Li, J.; Chen, H.Y.; Teng, Y.G.; Dong, Q.Q. Contamination characteristics and source apportionment of soil heavy metals in Lalin River basin. *Trans. Chin. Soc. Agric. Eng.* **2016**, *32*, 226–233.
52. Ali, M.H.; Mustafa, A.-R.A.; El-Sheikh, A.A. Geochemistry and spatial distribution of selected heavy metals in surface soil of Sohag, Egypt: A multivariate statistical and GIS approach. *Environ. Earth Sci.* **2016**, *75*, 1257. [CrossRef]
53. Gunawardena, J.; Ziyath, A.M.; Bostrom, T.E.; Bekessy, L.K.; Ayoko, G.A.; Egodawatta, P.; Goonetilleke, A. Characterisation of atmospheric deposited particles during a dust storm in urban areas of Eastern Australia. *Sci. Total Environ.* **2013**, *461–462*, 72–80. [CrossRef] [PubMed]

54. Zhou, J. Present situation and prospects of technologies for remediation of heavy-metal-contaminated soil around Jiangxi Guixi Smelter. *World Environ.* **2016**, *161*, 48–53.
55. Wu, S.X.; Zou, X.H.; Pan, X.L.; Cai, W.D.; Luo, Y.Q. Analysis and evaluation on heavy metal pollution condition of soil in farmland around Guixi Smelting Factory. *Jiangxi Sci.* **2012**, *30*, 779–783. [CrossRef]
56. Xu, S.; Gong, X.F.; Liu, C.Y.; Chen, C.L.; Zeng, H.Q.; Wang, J.J.; Li, Z.L.; Zheng, L. Ecological risk assessment and pollution analysis on the heavy metals contaminated soil around Guixi smeltery. *J. Nanchang Univ. (Nat. Sci.)* **2015**, *39*, 96–102. [CrossRef]
57. Zhang, Y.Z.; Yang, Z.H.; Qi, Y.H.; Zhang, X.T. Determination of Re, Mo and Se in copper smelting waste by indutively coupled plasma-atomic emission spectrometry. *Anhui Chem. Ind.* **2011**, *37*, 72–74.
58. Zhang, Z.Q.; Yuan, L.C.; Huang, L.Q.; Xu, Z.F. Trend and recovery of arsenic, antimony and bismuth in copper smelting. *Nonferrous Met. Sci. Eng.* **2019**, *10*, 13–19+27. [CrossRef]
59. Guan, Q.; Wang, F.; Xu, C.; Pan, N.; Lin, J.; Zhao, R.; Yang, Y.; Luo, H. Source apportionment of heavy metals in agricultural soil based on PMF: A case study in Hexi Corridor, northwest China. *Chemosphere* **2018**, *193*, 189–197. [CrossRef] [PubMed]
60. Liang, J.; Feng, C.; Zeng, G.; Gao, X.; Zhong, M.; Li, X.; Li, X.; He, X.; Fang, Y. Spatial distribution and source identification of heavy metals in surface soils in a typical coal mine city, Lianyuan, China. *Environ. Pollut.* **2017**, *225*, 681–690. [CrossRef] [PubMed]
61. Wang, M.; Li, S.S.; Li, X.Y.; Zhao, Z.Q.; Chen, S.B. An overview of current status of copper pollution in soil and remediation efforts in China. *Earth Sci. Front.* **2018**, *25*, 305–313. [CrossRef]
62. Meng, L.; Huang, T.H.; Chen, J.; Zhong, F.L.; Shi, J.C.; Xu, J.M. Safe utilization of farmland soil with cadmium pollution: Strategies and deliberations. *J. Zhejiang Univ. (Agric. Life Sci.)* **2019**, *45*, 263–271.
63. Xu, M.G.; Wu, H.W.; Liu, J. Evolution of heavy metal contents of three soils under long-term fertilizations. *J. Agro-Environ. Sci.* **2010**, *29*, 2319–2324.
64. Zhang, Y.J.; Liang, T.Z.; Feng, D.L.; Diao, L.Y.; Liang, M.Z. Research on trace element copper application in aquaculture. *J. Guangxi Agric.* **2019**, *34*, 57–62.
65. Cai, J.Y.; Zhang, W.L. Advances of epidemiological study on population exposure and health hazard of environmental cadmium pollution. *J. Environ. Hyg.* **2019**, *9*, 621–627. [CrossRef]
66. Deng, J.L.; Wang, Y.G.; Yu, D.; Li, X.J.; Yue, J.P. Effects of heavy metals on variation in bacterial communities in farmland soil of tailing dam collapse area. *Sci. Rep.* **2025**, *15*, 8100. [CrossRef]
67. Matse, D.T.; Geretharan, T.; van Gorp, E.F.; Anderson, S.; Jeyakumar, P.; Anderson, C.W.N. The Potential Impact of Long-Term Copper Fungicide Sprays on Soil Health in Avocado Orchards. *Environments* **2024**, *11*, 109. [CrossRef]

toxics

MDPI

Article

Screening of Profitable Chrysanthemums for the Phytoremediation of Cadmium-Contaminated Soils

Xinzhe Lu [1,2,*], Yanfang Chen [1,2], Jinqiu Song [3], Jiayu Bao [2,4], Chunzheng Dai [3], Rui Sun [1,2], Jiacheng Liu [2,4], Chenjiang Jin [5], Nanchong Zhong [3], Chunlei Huang [1,2,*] and Kokyo Oh [6]

1 Zhejiang Institute of Geosciences, Hangzhou 310007, China; 15268511668@126.com (Y.C.); nearysr@163.com (R.S.)
2 Technology Innovation Center of Ecological Evaluation and Remediation of Agricultural Land in Plain Area, MNR, Hangzhou 310007, China; 13069826196@163.com (J.B.); ljc13755921635@163.com (J.L.)
3 Zhejiang Zhezhong Geological Engineering Investigation Institute Co., Ltd., Jinhua 321001, China; sjq_840311@163.com (J.S.); 13757993951@139.com (C.D.); rockcold87@hotmail.com (N.Z.)
4 School of Earth Sciences and Resources, China University of Geosciences, Beijing 100083, China
5 Zhejiang Geology and Mineral Technology Co., Ltd., Hangzhou 311305, China; flcl@163.com
6 Center for Environmental Science in Saitama, 914 Kamitanadare, Kazo City 347-0115, Japan; o.kokyo@pref.saitama.lg.jp
* Correspondence: luxinzhe2016@163.com (X.L.); huangchunlei20202020@hotmail.com (C.H.)

Abstract: To explore the phytoremediation effect of ornamental chrysanthemums on cadmium (Cd)-contaminated farmland soil, a 2-year field trial was conducted on 23 chrysanthemum cultivars in Cd-contaminated soil in Zhejiang Province, China. The biomass yields, Cd content of the plants, Cd enrichment coefficient, and remediation efficiency were evaluated. The aboveground biomass of the tested chrysanthemums was 67.10–166.08 g/plant, the aboveground Cd content was 1.97–5.92 mg kg^{-1}, and the Cd enrichment coefficient was 2.98–9.84. In a screening test of twenty-three chrysanthemum cultivars, six cultivars, such as marigolds, were characterized by high cadmium accumulation, with the average cadmium accumulation of chrysanthemums exceeding 0.6 mg per plant, and the remediation of rhizosphere-contaminated soils took only 4–5 years. Fourteen chrysanthemum cultivars have good multiple-cropping characteristics, and five multiple-cropping chrysanthemum cultivars, such as QX-yz, have high heavy metal tolerance. The multiple-cropping JL-yg cultivars with higher Cd accumulation could be recommended for the remediation of Cd-contaminated farmland. The application of bamboo vinegar to the chrysanthemum rhizosphere effectively promoted Cd absorption. After estimating the economic benefits of artificially planting five dominant varieties of chrysanthemums for polluted farmland remediation, it is concluded that the annual income of a worker can be slightly higher than the average annual income level of local residents.

Keywords: phytoremediation; cadmium remove; soil contamination; chrysanthemum cultivars

1. Introduction

Heavy metal cadmium (Cd) is among the most serious and common pollutants in the soil. Soil pollution with heavy metals can harm animals and humans by getting into the food chain [1,2]. Currently, heavy metal-contaminated soil can be remediated by physical, chemical, or ecological methods (e.g., phytostabilization, microbial remediation, and biochar amendment) [3–7]. Phytoremediation employs hyperaccumulators to extract and concentrate soil heavy metals [8–11]. While phytoremediation can be cost effective when integrated with cash crops, traditional monoculture hyperaccumulators often lack economic returns [12]. In Chinese culture, chrysanthemum is known as one of the 'Four Gentlemen' and is widely used for ornamental purposes. A suburban area of Shanghai has been restored for safe use (Cd < 0.3 mg/kg, Zn < 150 mg/kg through the removal of excessive Cd and Zn by planting the chrysanthemum 'June White' [13,14]. Chrysanthemums have been found to have the advantages of easy cultivation, high survival rate, and heavy metal tolerance, and they have economic benefits as ornamental flowers, chrysanthemum tea, Chinese medicinal materials, etc. We consider the selection of chrysanthemums that bring economic benefits while achieving the reuse of contaminated land. The selection criteria are (1) BCF > 1.5; (2) market price ≥ RMB50/kg; and (3) biomass ≥ 3 tons/hectare.

In addition, some researchers are exploring the possibility of ethylenediamine tetraacetic acid (EDTA) [15], ethylenediamine succinic acid (EDDS) [16], diethyl triacetic acid (NTA) [17], and other synthetic chelating agents and low-molecular-weight organic acids to improve phytoremediation efficiency. Although a synthetic chelating agent can induce the desorption of heavy metal ions in the soil, such as Cd^{2+} and Pb^{2+}, it also creates the potential environmental problem of complexing and making other mineral elements in the soil unavailable, such as Ca^{2+} and Mg^{2+} [18], which is not conducive to the recovery and subsequent reuse of the remediated soil. Even at low doses, EDTA can inhibit the growth of rhizosphere microorganisms [19]. The biodegradable chelating agent NTA is not recommended for soil remediation as a result of its potential carcinogenicity. Citric acid can effectively promote plant absorption and the accumulation of Cd without affecting the normal growth of plants [20]. However, compared with an artificial chelating agent, a natural organic acid is more expensive, so there remain considerable limitations in the practical application and promotion of organic acids under the premise of limited remedial efficiency. The main component of bamboo vinegar is natural organic acid, which can increase the content of exchangeable heavy metals in the soil by forming soluble complexes with heavy metals or lowering the pH value of the soil to improve the efficiency of the plant's root system in absorbing heavy metals such as Cd^{2+}, Pb^{2+}, and other cations with similar ionic radii without affecting the normal growth of the plant. In addition, bamboo vinegar can be obtained in the process of bamboo charcoal production, and bamboo charcoal firing and product development have been industrialized.

In this study, heavy metal-contaminated farmland soil in a typical industrial area was selected for remediation. Chrysanthemum cultivars that have strong heavy metal accumulation, high tolerance characteristics, high biomass production, and additional economic benefits were selected for field trials. Testing the effectiveness of the chelating agents was combined with phytoremediation.

2. Materials and Methods

2.1. Study Site

This study was conducted in a 1 ha area of Cd-contaminated farmland (28°56′30″–28°56′50″ N, 120°07′30″–120°07′40″ E) located in a hardware-processing industrial town in Zhejiang Province, eastern China, with a subtropical monsoon climate. In 2020, a survey found that the Cd content in farmland soil from the study area exceeded the permitted limit of the Chinese national standard 'Soil Environmental Quality Standard for Soil Pollution Risk Control of Agricultural Land (Trial)' [13]. The content of Cd in soil is 0.32 mg kg^{-1}, and the exceeding rate of rice samples is nearly 30%. Cd pollution was mainly caused by sewage leaching from surrounding hardware-processing enterprises.

2.2. Experiment Design and Treatments

In the trial field, we established 23 test plots, and each plot covered an area of 12 m^2. The ridges with 0.3 m (height) multiplied by 0.3 m (width) were wrapped with film up to the bottom of the plow and encircled the test plots. Each plot had a one-way drainage ditch installed. A total of 23 cultivars of chrysanthemums were planted in a nearby plant factory for the experiment. Seedlings were grown between 10 March and 10 April 2021, and the best-growing cuttings were selected and transplanted to the experimental field. Each plot had only one type of chrysanthemum transplanted, and we planted the chrysanthemums in three rows with a spacing of 20 cm. The growth cycle of the chrysanthemum was from 10 April to 10 October. After the flowering period, all the chrysanthemums for the test were pruned, keeping a stem length of 10–12 cm and 3–4 chrysanthemum leaves, and the transplanting of chrysanthemums to the test field was not repeated in 2022. The 2022 campaign focused on multi-cropping effects rather than mere repetition, with sequential sampling conducted at the flowering stages of each crop cycle. Photographs of the test area and chrysanthemum cultivars are shown in Figure 1, and the full cultivar names are shown in Table 1. All the chrysanthemums purchased from Chrysanthemum Base of Nanjing Agricultural University, Jiangsu Province, China.

Figure 1. Chrysanthemum cultivars used in this study.

In order to improve the chrysanthemum remediation efficiency of contaminated soil, the *Chrysanthemum* × *morifolium* 'Qianxiuyinzhen' cultivar was selected to perform heavy metal chelator-coupled phytoremediation experiments based on the results of the 2021 experiment. There were five groups of treatments in the experiment, and each group of treatments had three experimental plots. The method and dosage of the chelating agent are shown in Table 2, all the chelating agents purchased from Jiangsu Fertilizer Industry

Co., LTD, Jiangsu Province, China. The chelating agent was sprayed between the roots of chrysanthemums in small quantities several times during the application.

Table 1. The full cultivar names of chrysanthemums.

Code	Variety	Code	Variety	Code	Variety
ZS-zc	Zhongshanzichen	QJ-bh1	Qingjianbaihuang1	JL-yg	Jinlingyuegui
L-xf	Lüxuanfeng	QJ-qn	Qingjianqingniao	ZS-hy	Zhongshanhongyun
QX-yz	Qianxiuyinzhen	JJ-hl	Jinjihongling	ZS-gh	Zhongshanguanghui
Marigold	Marigold	QJ-tp	Qingjiantaiping	JL-dc	Jinlingdianchun
QL-xh	Qunlongxihai	JS-hj	Jinsihuangju	JL-cj	Jinlingcuiju
QH-fmd	Qinhuaifenmudan	JS-bx	Jinsibaoxiao	JL-xy	Jinlingxiaoye
QJ-hg	Qingjianhuanggai	QJ-bh2	Qingjianbaihuang2	QJ-ht	Qingjianhuangtianzan
JH-hj	Jiuhuahongjing	QJ-dh	Qingjiandahuang		

Table 2. Experimental design to test the effects of bamboo vinegar and EDTA-Si application on Cd accumulation.

	Preparation	Application
Control (CK)	Purified water	1000 mL
Treatment 1	Diluted 15 mL of bamboo vinegar with purified water to 4500 mL	1500 mL
Treatment 2	Diluted 15 mL of bamboo vinegar with purified water to 3000 mL	1000 mL
Treatment 3	Dissolved 15 mg EDTA-Si in purified water and diluted it to 3000 mL	1000 mL
Treatment 4	Dissolved 15 mL of bamboo vinegar and 15 mg EDTA-Si in purified water and diluted it to 6000 mL	2000 mL

Note: EDTA-Si is a silica-immobilized EDTA composite.

2.3. Sampling and Experiment Analysis

Three groups of chrysanthemum plants and their rhizosphere soil were randomly taken from each plot during their flowering in 2021, and flower, stalk, and rhizosphere soil of multiple-crop chrysanthemum cultivars were collected in 2022. All samples were collected from the plant rhizosphere soil with a thickness of 0–20 cm, air-dried, ground, and passed through a 2 mm sieve before the test. The flower and stalk were dried, weighed, and pulverized for the test.

Cd content in soil and chrysanthemum plant samples was determined by ICP-MS (Agilent 7500 series, Agilent Technologies, Tokyo, Japan). Samples were pulverized (<100 mesh), digested with HNO_3-HF (3:1) in a microwave system (Mars 6, CEM), diluted to 50 mL with 2% HNO_3, and filtered (0.45 μm) prior to analysis, with detection limits of 0.03 and 0.0001. All sample analyses were strictly conducted in compliance with the standards [21,22], with quality control implemented using GBW series soil-certified reference materials (CRMs). The test results are of acceptable quality.

2.4. Evaluation of the Soil Remediation Capacity of Chrysanthemum Plants

Cadmium accumulation and years of remediation of contaminated soil were used in this study to assess the remediation capacity of chrysanthemum plants, and they were calculated as follows:

$$M_{soil} = S_{soil} \times h_{soil} \times \rho_{soil}/1000 \quad (1)$$

$$M_{Cd} = (Cd_{soil} - 0.3) \times M_{soil} \quad (2)$$

$$C_{Cd} = M_c \times Cd_c/1000 \quad (3)$$

$$T = M_{Cd}/C_{Cd} \quad (4)$$

where M_{soil} is the mass of the soil to be removed (kg); S_{soil} is the area of the soil to be removed (cm^2); h_{soil} is the depth of the soil to be removed (cm); ρ_{soil} is the density of the soil (g cm^{-3}); M_{Cd} is the mass of Cd in the soil to be removed (mg); Cd_{soil} is the Cd accumulation in the soil (mg); 0.3 is the current Chinese soil cadmium pollution control standard (mg kg^{-1}); C_{Cd} is the Cd accumulation of chrysanthemum plants (mg); M_c is the dry weight of chrysanthemum plants (g); Cd_c is the Cd content of chrysanthemum plants (mg kg^{-1}); and T is the remove period.

2.5. Data Analysis

This paper used Microsoft Office Excel 2019 for data reduction and analysis, the statistical software Statistical Product and Service Solutions 25 for statistical analysis, and Origin Pro 2022 for drawing statistical charts.

3. Results and Discussion

3.1. Cd Cumulative Abilities of 23 Chrysanthemum Cultivars

The Cd content of the rhizosphere soil ranged from 0.47 to 0.96 mg kg^{-1}, with an average of 0.65 mg kg^{-1}. Referring to the current Cd pollution control standard for soil in China (0.3 mg kg^{-1}), the exceedance rate of Cd in the soil samples was 100%. Plants absorb heavy metals through the roots and use transport proteins to translocate the metals to aboveground organs for storage. The goal of removing heavy metals in the soil can be achieved by harvesting the plants. Thus, high biomass production is an important factor for harvesting plants to extract greater quantities of heavy metals. All 23 tested chrysanthemum cultivars grew to the full flowering stage in the experimental field. As shown in Figure 2a, the aboveground partial biomass of the tested chrysanthemum cultivars ranged from 67.10 g to 166.08 g, with an average of 111.25 g. A marked difference can be observed between the highest and the lowest values. Among the tested cultivars, the six cultivars with biomass exceeding 140 g were QJ-ht, marigolds, JL-dc, ZS-zc, ZS-gh, and ZS-hy, in that order. The comparison of Cd content between chrysanthemum plants and rhizosphere soil can reflect the Cd absorption capacity of chrysanthemum plants. The aboveground Cd contents of the tested chrysanthemum cultivars ranged from 1.97 to 5.92 mg kg^{-1}, with an average of 4.45 mg kg^{-1}. As shown in Figure 2b, chrysanthemum plants had a high enrichment capacity for Cd, with enrichment coefficients ranging from 2.98 to 9.84 and averaging up to 6.95. The chrysanthemum cultivars with enrichment coefficients over 8 were, in descending order, QJ-hg, QJ-dh, JL-yg, JL-cj, ZS-gh, QJ-bh1, marigolds, and JS-hj. Under natural conditions in the field, the ability of chrysanthemums in this experiment to accumulate soil heavy metal Cd was higher than reported in high-accumulation plants such as *Arabis paniculata* [20], *Centella Asiatica* [23], *Phytolacca americana* [24], *Potentilla griffithii* [25], *Sigesbeckia orientalis* [26], and *Viola baoshanensis* [27].

The mass of Cd that can be accumulated by each chrysanthemum plant per season is determined by biomass and Cd content [28], and the mass of Cd accumulated by a single chrysanthemum plant in a single season is shown in Figure 2c. The mass of Cd accumulated by the 23 chrysanthemum cultivars ranged from 0.26 to 0.83 mg, with an average of 0.49 mg (Equation (3)). The six chrysanthemum cultivars that accumulated over 0.6 mg Cd were, in descending order, marigold, ZS-gh, QJ-ht, JL-cj, QJ-hg, and JL-dc. Based on the 20 cm chrysanthemum spacing, the average area occupied by each chrysanthemum was 400 cm^2, h_{soil} was taken as 20 cm, and ρ_{soil} was 1.1 g cm^{-3}. As shown in Figure 2d, the remediation of rhizosphere Cd-contaminated soil by the test chrysanthemums required 4–13 stages, with an average of 8 stages (Equations (1), (2), and (4)). There were nine chrysanthemum cultivars with five or fewer remediation stages: ZS-gh, JL-dc, JL-cj, marigolds, QJ-ht, QJ-hg, ZS-hy, QL-xh, and QJ-dh.

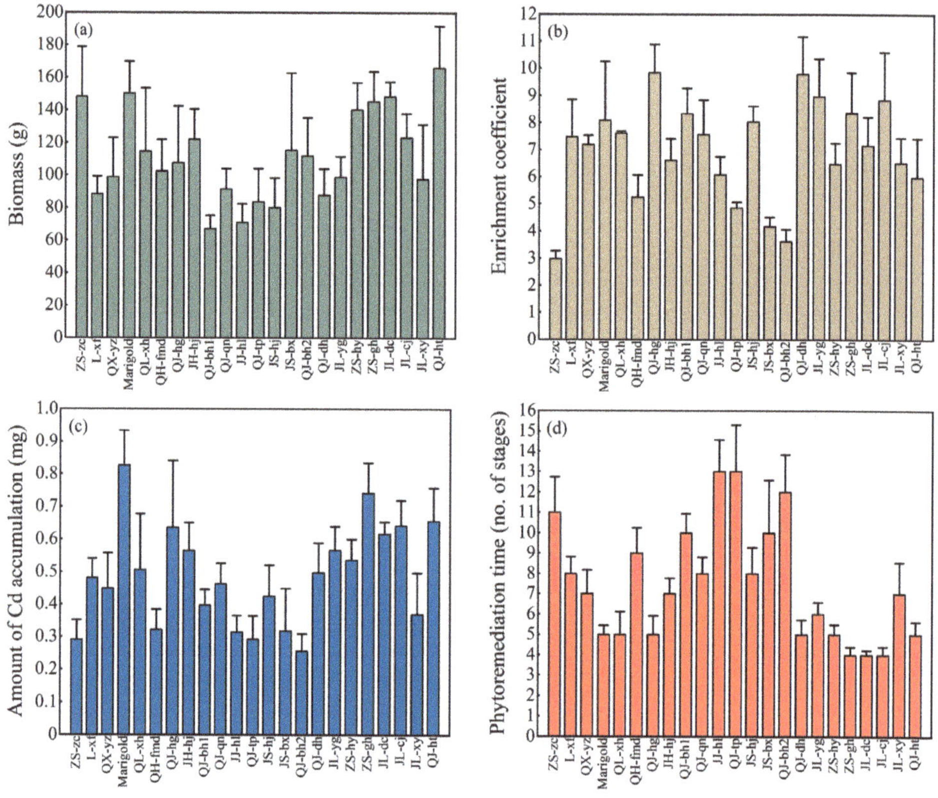

Figure 2. Aboveground biomass (**a**), enrichment coefficient (**b**), Cd accumulation (**c**), and remediation stage (**d**) of 23 chrysanthemum cultivars.

The 23 tested chrysanthemum cultivars were compared with domestic and foreign reported Cd hyperaccumulating plants, such as *Portulaca oleracea* [29] and *Sedum alfredii* [30], and a variety of chrysanthemums in this test had obvious Cd hyperaccumulation characteristics and had excellent efficiency in removing soil Cd.

3.2. Heavy Metal Tolerance in Multiple-Cropping Chrysanthemum Cultivars

Multiple-cropping chrysanthemum cultivars can reduce the cost and increase the efficiency of soil remediation. Fourteen chrysanthemum cultivars grew to full bloom after overwintering in 2022: ZS-zc, L-xf, QX-yz, QJ-hg, JH-hj, QJ-tp, QJ-bh2, QJ-dh, JL-yg, ZS-gh, JL-dc, JL-cj, JL-xy, and QJ-ht. The tolerance of these chrysanthemum cultivars to heavy metals is a prerequisite for their continued use in soil remediation, as shown in Figure 3, which compares the biomass of multiple-cropping chrysanthemums in 2021–2022. The biomass of five chrysanthemum cultivars, QX-yz, QJ-hg, QJ-bh2, JL-xy, and JL-yg, remained stable compared to the previous year, showing that these chrysanthemum cultivars are more tolerant to heavy metals. Most of the multiple-cropping chrysanthemum biomass was lower than that of the previous year, which may be due to the decline in soil fertility or the growth inhibitory effect of the heavy metal Cd on chrysanthemum plants. Based on Cd content in chrysanthemum plants in 2021, the remediation of rhizosphere soils should be extended by an average of 3.29 stages.

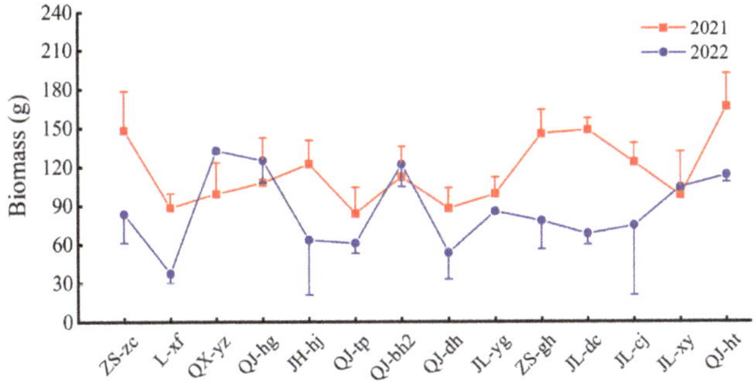

Figure 3. Changes in aboveground biomass of multiple-cropping chrysanthemum cultivars.

3.3. Enrichment Coefficient to Different Aboveground Parts of Multiple-Cropping Chrysanthemum Cultivars

The mechanism of heavy metal absorption may differ among plant cultivars growing in heavy metal-contaminated soil. Some toxic heavy metals are fixed in the root, whereas other cultivars tend to transport the heavy metal to the aboveground parts [31]. The xylem vessels play an important role in the long-distance transport of solutes from the roots, and some toxic heavy metals in the roots can bypass the phloem to enter the xylem and are then transported to aboveground organs [32]. Previous studies have reported that the relative Cd accumulation factors for different organs of chrysanthemums were ranked as follows: leaf > stem > flower [33]. In the present study, 14 multiple-cropping chrysanthemums were analyzed for the content and enrichment coefficient of Cd in flowers and stalk, as shown in Figure 4. The Cd content in the flower of 14 chrysanthemum cultivars was distributed between 1.98 and 4.98 mg kg^{-1}, with an average of 3.28 mg kg^{-1}, and the enrichment coefficients were distributed between 2.67 and 8.94, with an average of 5.30. The Cd content of the stalk was distributed between 4.20 and 15.00 mg kg^{-1} with an average of 8.19 mg kg^{-1}, and the enrichment coefficients were distributed between 7.80 and 23.77, with an average of 12.86. The Cd content and enrichment coefficients of all tested chrysanthemum stalks were higher than those of flowers, and chrysanthemum plant stalks were superior to flowers, which showed that chrysanthemum stalks accounted for the major portion of the whole chrysanthemum plants in terms of Cd accumulation.

3.4. Cd Enrichment Efficiency of Chrysanthemum Plants Under Organic Acid and EDTA-Si

The QX-yz cultivar with stable biomass and high Cd accumulation was subjected to bamboo vinegar and EDTA-Si application tests, as shown in Figure 5. There was no significant change ($p < 0.05$) in chrysanthemum biomass in treatments 1 and 2 compared to the control, which showed that the concentration of bamboo vinegar used in the experiment did not have a toxic effect on chrysanthemum growth. The Cd content of chrysanthemum plants in treatments 1 and 2 was significantly increased by 29.52 and 16.65 percentage points ($p < 0.05$) compared with the control, which indicated that the bamboo vinegar promoted Cd^{2+} accumulation through the roots of chrysanthemum plants, and the formation of Cd^{2+} complexes and the decrease in soil pH led to the increase in the Cd content of the exchangeable state in the soil. Cd accumulation in chrysanthemum plants was elevated by 29.63 and 34.99 percentage points in treatments 1 and 2 compared with the control. Treatment 3 chrysanthemum biomass significantly decreased by 62.22 percentage points ($p < 0.05$) compared to the control; treatment 4 significantly decreased by 74.85 and 77.53 percentage points ($p < 0.05$) compared to the control and treatment 2, which showed that the growth of chrysanthemum plants was inhibited. Similarly, the toxic effect of EDTA on plants was observed in the enrichment study of heavy metals in *Brassica juncea* [34].

Studies have shown that high levels of Cd are toxic to chrysanthemum plants, and in addition, EDTA may complex Ca^{2+} and Zn^{2+} to disrupt the biofilm, causing damage to chrysanthemum plants during the growth period [35]. Since the chelation of EDTA-Si with Cd^{2+} in the soil formed a stable water-soluble complex, the Cd content of chrysanthemum plants in treatments 3 and 4 significantly increased by 97.53 and 61.02 percentage points ($p < 0.05$). A comparison of treatment 2 with treatment 3 showed that the biological activity of Cd under EDTA-Si treatment was higher than that of bamboo vinegar ($p < 0.05$). However, the reduced biomass of chrysanthemum plants under EDTA-Si treatment caused a decrease in Cd accumulation in chrysanthemum plants in treatments 3 and 4.

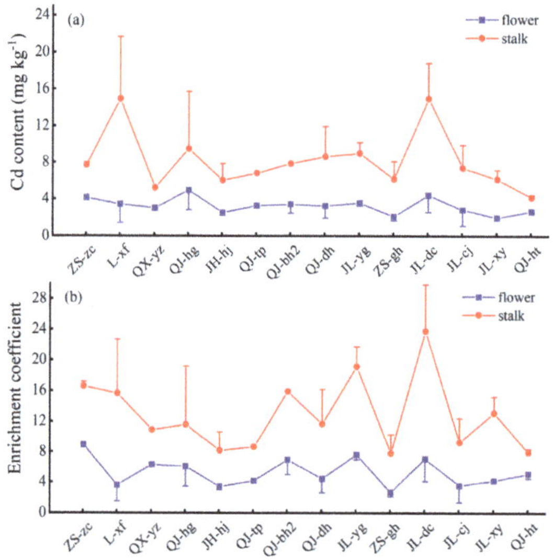

Figure 4. The Cd content (**a**) and enrichment coefficient (**b**) in flowers and stalks.

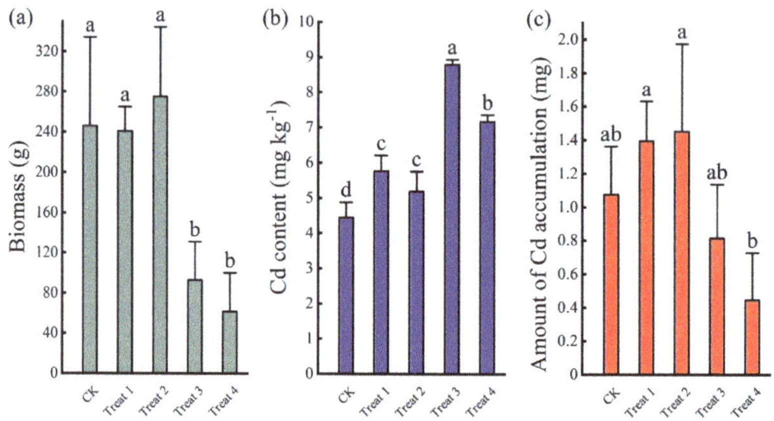

Figure 5. Aboveground biomass (**a**), Cd content (**b**), and Cd accumulation (**c**) of chrysanthemum plants under organic acid and EDTA-Si. Note: Lowercase letters a,b,c,d above the error bar represents the significant differences ($p < 0.05$) among different treatments, these containing the same letters show no significant difference (i.e., ab and a), while those with different letters indicate a significant difference between the two (i.e., a and b).

Bamboo vinegar solution and EDTA-Si could significantly promote the accumulation of Cd in chrysanthemum plants in this experiment; in addition, the use of EDTA-Si caused a significant decrease in the biomass of chrysanthemums; however, the effects on chrysanthemum plants at different application rates were not clear. Although treatments 1 and 2 used different concentrations of bamboo vinegar, they did not differentiate in the amount applied, and no significant changes in chrysanthemum biomass and Cd content were observed ($p < 0.05$). Therefore, the rational application of bamboo vinegar and EDTA-Si to chrysanthemum plants needs to be further investigated. As far as the results are concerned, treatments 1 and 2 better improved the Cd accumulation efficiency of chrysanthemum plants, and under the premise of constant biomass, the remediation of contaminated soil by QX-yz cultivars after applying bamboo vinegar could be advanced by 1 year, which can be used for the remediation of contaminated soil by popularizing the use of chrysanthemum plants.

3.5. Estimation of the Economic Benefits of Chrysanthemum Remediation for Cd-Contaminated Agricultural Land

The remediation of heavy metal-contaminated soil through engineering excavation methods often requires a long time and high costs [36]. Moreover, during the remediation process, the contaminated farmland generates no economic returns. Although phytoremediation has received increasing attention in recent decades as an environmentally friendly and cost-effective green technology, its practical application remains limited. This is primarily due to the slow growth and low biomass of remediating plants, resulting in low remediation efficiency, and the long duration typically required for the remediation of heavy metal-contaminated farmland. A more critical issue is that most hyperaccumulator plants for heavy metals are non-economic species, meaning that the long-term remediation process of contaminated farmland does not generate income but requires significant investment [37]. To promote the practical application of phytoremediation, it is necessary to establish an effective and economically beneficial phytoremediation system. This would enable owners of contaminated farmland to earn annual economic returns during the phytoremediation of contaminated soil.

In this study, the selected chrysanthemum varieties for remediation can be continuously harvested and sold during the remediation process of contaminated soil, generating economic benefits. If we consider only the input costs and revenue from chrysanthemum sales, without considering the environmental and social benefits of contaminated soil remediation, we have calculated a detailed economic account.

To enhance the effectiveness of phytoremediation in removing heavy metals from the soil, it is necessary to construct some auxiliary planting facilities to ensure double cropping of chrysanthemums and maximize their biomass. The construction and maintenance costs of chrysanthemum planting facilities are 0.75×10^4 RMB per hectare per year, while the costs of fertilizers and pesticides for chrysanthemum cultivation are 3.15×10^4 RMB per hectare per year, and the costs of machinery and tools are 0.525×10^4 RMB per hectare per year. Additionally, land rental costs are 1.8×10^4 RMB per hectare per year.

The average yield of chrysanthemum branches per hectare of contaminated farmland is 3.89×10^6 branches (Table 3). Most chrysanthemums are sold as ceremonial items to funeral homes across Zhejiang Province. The price of chrysanthemums varies with time and quality. Generally, ordinary-grade chrysanthemums account for 80%, with each bundle (14 branches) selling for 2.5 RMB; high-grade chrysanthemums account for 5%, with each bundle selling for 5 RMB; and low-grade white chrysanthemums account for 15%, with each bundle selling for 1 RMB.

Table 3. Annual yield of chrysanthemum flowers in the contaminated farmland and theoretical annual income per hectare.

Chrysanthemum Varieties	Flower Count (hm^{-2} $year^{-1}$)	Net Income (RMB hm^{-2} $year^{-1}$)
QX-yz	$3.96 \times 10^6 \pm 0.42 \times 10^6$	$61.67 \times 10^4 \pm 9.26 \times 10^4$
QJ-hg	$3.89 \times 10^6 \pm 0.72 \times 10^6$	$60.46 \times 10^4 \pm 8.38 \times 10^4$
QJ-bh2	$4.02 \times 10^6 \pm 0.86 \times 10^6$	$62.69 \times 10^4 \pm 11.34 \times 10^4$
JL-xy	$3.69 \times 10^6 \pm 0.41 \times 10^6$	$57.03 \times 10^4 \pm 7.87 \times 10^4$
JL-yg	$4.16 \times 10^6 \pm 0.58 \times 10^6$	$65.09 \times 10^4 \pm 10.43 \times 10^4$
Average	$5.31 \times 10^6 \pm 1.00 \times 10^6$	$60.46 \times 10^4 \pm 8.63 \times 10^4$
Marigold		16.88×10^4~55.70×10^4

A farmer can plant chrysanthemums on 0.1 hectares of land and harvest two crops per year. After deducting all the aforementioned costs, the farmer's net income from chrysanthemum cultivation is 6.17×10^4 RMB, which is higher than the average annual income of residents in Yongkang City, Zhejiang Province, which is 5.67×10^4 RMB.

This indicates that by planting chrysanthemums for the remediation of heavy metal-contaminated soil, there is no need to pay for land clearing costs and land idleness losses, and economic benefits can be earned through the sale of chrysanthemums. The net income from chrysanthemum cultivation in this study is higher than that generated by the phytoremediation of heavy metal-contaminated soil using marigolds [38]. In conclusion, the selected dominant chrysanthemum varieties exhibit vigorous growth, high biomass, strong tolerance to toxic heavy metals, and demonstrate strong Cd accumulation and removal effects. This study primarily focuses on the accumulation and removal effects of chrysanthemums on Cd, and they may have similar effects on other harmful heavy metals. More importantly, planting white chrysanthemums for phytoremediation can also generate considerable economic benefits.

4. Conclusions

In this study, 23 chrysanthemum cultivars were tested. Six Cd hyperaccumulating chrysanthemum cultivars were marigolds, ZS-gh, QJ-ht, JL-cj, QJ-hg, and JL-dc, the average Cd accumulation per plant was over 0.6 mg, and their remediation of rhizosphere soils required 4–5 years. There are 14 multiple-cropping cultivars, of which the biomass of chrysanthemum cultivars QX-yz, QJ-hg, QJ-bh2, JL-xy, and JL-yg remain stable. The highest Cd accumulation was 0.64 mg in the QJ-hg single plant. The use of EDTA-Si can bring about toxic effects and cause chrysanthemum biomass to decline. Bamboo vinegar is more helpful to increase the Cd accumulation of chrysanthemum plants. In this experiment, the application of bamboo vinegar led to Cd accumulation within chrysanthemums being elevated by up to 34.99 percentage points, and the number of stages of soil remediation was expected to be reduced by 1. Additionally, chrysanthemum sales provide sustainable economic benefits to the phytoremediation process, with practitioners in chrysanthemum cultivation earning above the local residents' average annual income level.

Author Contributions: X.L.: conceptualization, formal analysis, funding acquisition, investigation, methodology, project administration, supervision, visualization, writing—original draft, and writing—review and editing. Y.C.: investigation, writing—original draft, and writing—review and editing. J.S.: investigation, writing—original draft, and writing—review and editing. J.B.: investigation and writing—original draft. C.D.: investigation and writing—review and editing. R.S.: investigation and writing—review and editing. J.L.: investigation and writing—review and editing. C.J.: investigation and writing—review and editing. N.Z.: investigation and writing—review and editing. C.H.: funding acquisition, investigation, and writing—review and editing. K.O.: funding acquisition, investigation, and writing—review and editing. All authors have read and agreed to the published version of the manuscript.

Funding: This work was supported by the Science and Technology Programs of the Department of Natural Resources of Zhejiang Province, China (Grant Nos. 2020006 and 2023021 and No. 2020-45); the Ecological Environment Scientific Research Achievement Promotion Program of Zhejiang Province (Grant No. 2021XM0040); the Science and Technology Project of Jinhua City (2022-3-065); and the Japan Society for the Promotion of Science Research Foundation (No. 16H0533).

Institutional Review Board Statement: Not applicable.

Informed Consent Statement: Not applicable.

Data Availability Statement: Data will be made available upon request.

Conflicts of Interest: Authors Jinqiu Song, Chunzheng Dai and Nanchong Zhong were employed by the company Zhejiang Zhezhong Geological Engineering Investigation Institute Co., Ltd. Author Chenjiang Jin was employed by the company Zhejiang Geology and Mineral Technology Co., Ltd. The re-maining authors declare that the research was conducted in the absence of any commercial or fi-nancial relationships that could be construed as a potential conflict of interest.

References

1. Wang, G.; Zhang, Q.; Du, W.; Ai, F.; Yin, Y.; Ji, R.; Guo, H. Microbial communities in the rhizosphere of different willow genotypes affect phytoremediation potential in Cd contaminated soil. *Sci. Total Environ.* **2021**, *769*, 145224. [CrossRef]
2. Xia, X.; Yang, Z.; Yu, T.; Zhang, C.; Hou, Q. Predicting spatial and temporal variation of Cd concentration in rice grains in the Lower Changjiang Plain during 2004–2014 based on soil geochemical survey data with GIS. *J. Geochem. Explor.* **2018**, *200*, 276–283. [CrossRef]
3. Arteaga, J.F.; Gluhar, S.; Kaurin, A.; Lestan, D. Simultaneous removal of arsenic and toxic metals from contaminated soil: Laboratory development and pilot scale demonstration. *Environ. Pollut.* **2022**, *294*, 118656. [CrossRef]
4. Kalyvas, G.; Bilias, F.; Gasparatos, D.; Zafeiriou, I.; Eissa, R.; Karamountzou, E.; Massas, I. Enhanced As, Pb and Zn uptake by Helianthus annuus from a Heavily contaminated mining soil amended with edta and olive mill wastewater due to increased element mobilization, as verified by sequential extraction schemes. *Environments* **2022**, *9*, 61. [CrossRef]
5. Liu, L.; Li, W.; Song, W.; Guo, M. Remediation techniques for heavy metal-contaminated soils: Principles and applicability. *Sci. Total Environ.* **2018**, *633*, 206–219. [CrossRef] [PubMed]
6. Liu, W.; Wei, D.; Mi, J.; Shen, Y.; Cui, B.; Han, C. Immobilization of Cu(II) and Zn(II) in simulated polluted soil using sulfurizing agent. *Chem. Eng. J.* **2015**, *277*, 312–317. [CrossRef]
7. Vaxevanidou, K.; Papassiopi, N.; Paspaliaris, I. Removal of heavy metals and arsenic from contaminated soils using bioremediation and chelant extraction techniques. *Chemosphere* **2008**, *70*, 1329–1337. [CrossRef]
8. Gardea-Torresdey, J.; Peralta-Videa, J.; de la Rosa, G.; Parsons, J. Phytoremediation of heavy metals and study of the metal coordination by X-ray absorption spectroscopy. *Coord. Chem. Rev.* **2005**, *249*, 1797–1810. [CrossRef]
9. Kafle, A.; Timilsina, A.; Gautam, A.; Adhikari, K.; Bhattarai, A.; Aryal, N. Phytoremediation: Mechanisms, plant selection and enhancement by natural and synthetic agents. *Environ. Adv.* **2022**, *8*, 100203. [CrossRef]
10. Sarwar, N.; Imran, M.; Shaheen, M.R.; Ishaque, W.; Kamran, M.A.; Matloob, A.; Rehim, A.; Hussain, S. Phytoremediation strategies for soils contaminated with heavy metals: Modifications and future perspectives. *Chemosphere* **2017**, *171*, 710–721. [CrossRef]
11. Solomou, A.D.; Germani, R.; Proutsos, N.; Petropoulou, M.; Koutroumpilas, P.; Galanis, C.; Maroulis, G.; Kolimenakis, A. Utilizing mediterranean plants to remove contaminants from the soil environment: A short review. *Agriculture* **2022**, *12*, 238. [CrossRef]
12. Wang, J.; Delavar, M.A. Techno-economic analysis of phytoremediation: A strategic rethinking. *Sci. Total Environ.* **2023**, 165949. [CrossRef]
13. *GB 15618-2018*; Soil Environmental Quality Risk Control Standard for Soil Contamination of Agricultural Land (Trial). Ministry of Ecology and Environment of People's Republic of China: Beijing, China, 2018.
14. Luo, F.; Hu, X.-F.; Oh, K.; Yan, L.-J.; Lu, X.-Z.; Zhang, W.-J.; Yonekura, T.; Yonemochi, S.; Isobe, Y. Using profitable chrysanthemums for phytoremediation of Cd-and Zn-contaminated soils in the suburb of Shanghai. *J. Soils Sediments* **2020**, *20*, 4011–4022. [CrossRef]
15. Beiyuan, J.; Lau, A.Y.T.; Tsang, D.C.W.; Zhang, W.; Kao, C.-M.; Baek, K.; Ok, Y.S.; Li, X.-D. Chelant-enhanced washing of CCA-contaminated soil: Coupled with selective dissolution or soil stabilization. *Sci. Total Environ.* **2018**, *612*, 1463–1472. [CrossRef]

16. Fabbricino, M.; Ferraro, A.; Luongo, V.; Pontoni, L.; Race, M. Soil Washing Optimization, Recycling of the Solution, and Ecotoxicity Assessment for the Remediation of Pb-Contaminated Sites Using EDDS. *Sustainability* **2018**, *10*, 636. [CrossRef]
17. Jani, Y.; Hogland, W. Chemical extraction of trace elements from hazardous fine fraction at an old glasswork dump. *Chemosphere* **2017**, *195*, 825–830. [CrossRef]
18. Zhang, S.; Xu, X.; Zhong, Q.; Zhang, C.; Jia, Y.; Li, T.; Deng, O.; Li, Y. Heavy metal removal by GLDA washing: Optimization, redistribution, recycling, and changes in soil fertility. *Sci. Total Environ.* **2016**, *569–570*, 557–568. [CrossRef]
19. López-Rayo, S.; Sanchis, I.; Ferreira, C.; Lucena, J. [S,S]-EDDS/Fe: A new chelate for the environmentally sustainable correction of iron chlorosis in calcareous soil. *Sci. Total Environ.* **2018**, *647*, 1508–1517. [CrossRef]
20. Sinhal, V.; Srivastava, A.; Singh, V. EDTA and citric acid mediated phytoextraction of Zn, Cu, Pb and Cd through marigold (*Tagetes erecta*). *J. Environ. Biol. / Acad. Environ. Biol.* **2010**, *31*, 255–259.
21. *DD 2005-03*; Technical Requirements for Sample Analysis in Ecological Geochemical Evaluation (Trial). China Geological Survey: Beijing, China, 2005.
22. *DZ/T 0279-2016*; Analysis Methods for Regional Geochemical Sample. Ministry of Land and Resources of China: Beijing, China, 2016.
23. Hu, P.; Zhou, X.; Qiu, R.; Tang, Y.; Ying, R. Cadmium tolerance and accumulation features of Zn-hyperaccumulator *Potentilla griffithii* var. velutina. *J. Agro-Environ. Sci.* **2007**, *26*, 2221–2224.
24. Liu, K.; Zhou, Z.; Yu, F.; Chen, M.; Chen, C.; Zhu, J.; Jiang, Y. A newly found cadmium hyperaccumulator-centella asiatica linn. *Environ. Bull.* **2016**, 2668.
25. Liu, W.; Shu, W.; Lan, C. Viola baoshanensis, a plant that hyperaccumulates cadmium. *Chin. Sci. Bull.* **2004**, *49*, 29–32. [CrossRef]
26. Liu, X.; Peng, K.; Wang, A.; Lian, C.; Shen, Z. Cadmium accumulation and distribution in populations of *Phytolacca americana* L. and the role of transpiration. *Chemosphere* **2010**, *78*, 1136–1141. [CrossRef] [PubMed]
27. Tang, Y.; Qiu, R.; Zeng, X.; Ying, R.; Yu, F.; Zhou, X. Lead, zinc, cadmium hyperaccumulation and growth stimulation in Arabis paniculata Franch. *Environ. Exp. Bot.* **2009**, *66*, 126–134. [CrossRef]
28. Zhang, S.; Lin, H.; Deng, L.; Gong, G.; Jia, Y.; Xu, X.; Li, T.; Li, Y.; Chen, H. Cadmium tolerance and accumulation characteristics of *Siegesbeckia orientalis* L. *Ecol. Eng.* **2013**, *51*, 133–139. [CrossRef]
29. Li, J.; Liao, B.; Lan, C.; Ye, Z.; Baker, A.; Shu, W. Cadmium Tolerance and Accumulation in Cultivars of a High-Biomass Tropical Tree (*Averrhoa carambola*) and Its Potential for Phytoextraction. *Agric. Water Manag.* **2010**, *39*, 1262–1268. [CrossRef]
30. Tiwari, K.; Dwivedi, S.; Mishra, S.; Srivastava, S.; Tripathi, R.; Singh, N.; Chakraborty, S. Phytoremediation efficiency of *Portulaca tuberosa* rox and *Portulaca oleracea* L. naturally growing in an industrial effluent irrigated area in Vadodra, Gujrat, India. *Environ. Monit. Assess.* **2008**, *147*, 15–22. [CrossRef]
31. Yang, X.; Long, X.; Ye, H.; He, Z.; Calvert, D.V.; Stoffella, P.J. Cadmium Tolerance and Hyperaccumulation in a New Zn-Hyperaccumulating Plant Species (*Sedum alfredii* Hance). *Plant Soil* **2004**, *259*, 181–189. [CrossRef]
32. Pulford, I.; Watson, C. Phytoremediation of Heavy Metal-Contaminated Land by Trees—A Review. *Environ. Int.* **2003**, *29*, 529–540. [CrossRef]
33. Seregin, I.; Kozhevnikova, A. Roles of root and shoot tissues in transport and accumulation of cadmium, lead, nickel, and strontium. *Russ. J. Plant Physiol.* **2008**, *55*, 1–22. [CrossRef]
34. Lal, K.; Minhas, P.S.; Chaturvedi, R.K.; Yadav, R. Extraction of cadmium and tolerance of three annual cut flowers on Cd-contaminated soils. *Bioresour. Technol.* **2008**, *99*, 1006–1011. [CrossRef] [PubMed]
35. Blaylock, M.J.; Salt, D.E.; Dushenkov, S.; Zakharova, O.; Gussman, C.; Kapulnik, Y.; Ensley, B.D.; Raskin, I.J. Enhanced accumulation of Pb in Indian mustard by soil-applied chelating agents. *Environ. Sci. Technol.* **1997**, *31*, 860–865. [CrossRef]
36. Pasternak, C. A novel role of Ca^{2+} and Zn^{2+}: Protection of cells against membrane damage. *Biosci. Rep.* **1988**, *8*, 579–583. [CrossRef]
37. Gong, Y.; Zhao, D.; Wang, Q.L. An overview of field-scale studies on remediation of soil contaminated with heavy metals and metalloids: Technical progress over the last decade. *Water Res.* **2018**, *147*, 440–460. [CrossRef]
38. Chintakovid, W.; Visoottiviseth, P.; Khokiattiwong, S.; Lauengsuchonkul, S. Potential of the hybrid marigolds for arsenic phytoremediation and income generation of remediators in RonPhibun District, Thailand. *Chemosphere* **2008**, *70*, 1532–1537. [CrossRef]

MDPI AG
Grosspeteranlage 5
4052 Basel
Switzerland
Tel.: +41 61 683 77 34

Toxics Editorial Office
E-mail: toxics@mdpi.com
www.mdpi.com/journal/toxics

www.ingramcontent.com/pod-product-compliance
Lightning Source LLC
Chambersburg PA
CBHW062056220926
51850CB00010B/2790
9783725879854